# Engineering Colorado

# Engineering Colorado

History of Consulting Engineers Council of Colorado
**by Richard Weingardt and Fu Hua Chen**

*Foreword by Governor Roy Romer*

A Jacqueline Enterprises Publication
Englewood, Colorado USA 80111

# Table of Contents

***IN MEMORIUM***
*Rhuel Andersen*
*E.H. Carter*
*Vern Konkel*
*George Koonsman*
*Ken Murray*
*Alvin Swanson*
*John Tracy*

# Foreword

## By The Honorable Roy Romer
Governor, State of Colorado

Engineers have played an important part in the development of Colorado. From the early days when the gold rush brought our first permanent settlements, engineers have been involved in designing highways, railroads, tunnels and canals over and through our mountains and along our rivers.

Modern day consulting engineers plan the infrastructure of roads, bridges, airports, reservoirs, water treatment plants, waste facilities and public and private buildings.

Many human needs can be met by an engineer's creative application of scientific knowledge. Engineers can improve the quality of life by using expertise and ingenuity to solve practical engineering problems.

I am pleased that the Consulting Engineers Council of Colorado has contributed to the history and growth of our great state.

Roy Romer
Governor

*Governor Roy Romer presents his Proclamation designating February 19-25, 1989 as "Engineers Week" to Colorado engineers. Present at the signing ceremonies were (left to right) Romer, Orville Stoddard, president-elect of Professional Engineers of Colorado; Paul Cheng, president of Colorado Engineers Council; Richard Weingardt, president of Consulting Engineers Council of Colorado. (Photography by Claude Powe)*

*The Proclamation reads: WHEREAS, necessities and ideas guide the thoughts of dreamers to provide for the convenience of society; and WHEREAS the leadership, and professional and practical experience of engineers help develop these ideas and needs into reality; and WHEREAS, the nation will recognize the continuing contributions of America's engineers by observing Engineers Week with the motto, "Engineers: Turning Ideas into Reality";*

*NOW, THEREFORE, I, Roy Romer, Governor of Colorado, proclaim February 19-25, 1989, as ENGINEERS WEEK in the State of Colorado.*

# Introduction

This book is about the evolution and the shaping of Consulting Engineers Council of Colorado (CECC). It is the story of the growth of the consulting industry in the state and the impact consulting engineers in the association have had engineering construction projects in Colorado.

Recognizing the trust the public has for the profession of engineering, consulting engineers organized CECC to: (1) improve the private practice of consulting engineering, (2) protect public safety and welfare, (3) further business relations in the community, (4) maintain high professional standards and (5) exchange business experiences and cordial relations among its members. It is a non-profit corporation.

Member firms range from the smallest to the largest firms, from the oldest to the newest, and they are experts in a wide range of disciplines. All of them share a concern for providing the very best engineering services possible. CECC firms are evaluated for membership by their peers and are committed to the "code of ethics" which is part of this publication.

Founded in 1954-55, CECC this year celebrates 34 years of service to consulting engineering in Colorado. Members of the organization have had a profound impact on shaping the state as well as the engineering profession.

The founding of the Consulting Engineers Association grew out of the need of owners and managers of consulting firms to deal with business practices.

According to Orley Phillips, one of the major reasons for establishing CECC was that NSPE (National Society of Professional Engineers) and other engineering groups were not adequately dealing with business issues affecting private consultants. Paul Koch, another founding member, reminisces that early organizational meetings were held at the home of Francis Stark and one of the biggest problems for consulting mechanical engineers was "manufacturer's reps. (and salesmen) doing free engineering." Both Koch and Stark, who was a business associate of James Konkel, were mechanical engineers. Phillips is a civil-structural engineer.

To put the beginning of CECC in perspective, it is appropriate to remember that in 1955 Ike (President Eisenhower) had just increased the nation's minimum wage to $1 per hour. Orley Phillips recalls that "high pay" for a principal in a consulting engineering firm would have been $750 per month.

Disneyland, the super amusement park creation of Walt Disney, a steel and concrete "Never Never Land," opened at a cost of $17 million.

"Big Ed," Edwin Johnson, the Democrat's "grand old man," was governor of Colorado. The state's total population was around one and one half million people and the new Air Force Academy near Colorado Springs had received its first enrollment of 306 cadets.

Sources of our research included CECC publications, newspapers, personal papers and interviews. Many have contributed their efforts and knowledge to the evolution of this history.

The authors acknowledge our indebtedness to many CECC members, especially those who were around in the early years. We appreciate the generous sharing of their historical documents, photographs,

10

and their recollections of "what it was like."

We are grateful to Ed Leland, photographer, for use of his photograph of Maroon Bells mountain for our cover; Elaine Langeberg and Wendy Elmore for typing the manuscript, *and Elaine for her tenacity in getting information from past presidents for their profiles.* We also thank Sandy Donnel and Jane Wales for proofreading; *all four of them for correcting our imaginative spelling*; and David Weingardt for typesetting, compiling the index, and book design.

The writing of history calls for the sifting of sources and memories in an attempt to create a narrative that captures the vital spirit of what happened between then and now. Our hope, along with those who helped, is that our attempt to do this for the history of CECC will bring you pleasure today and tomorrow.

*Richard Weingardt*
*Fu Hua Chen*

*Kenneth Murray, 1955, first president of Consulting Engineers (Association) of Colorado.*

*The tenure of current CECC presidents runs from spring to spring; the term, for instance, of the 1970 president would actually be from spring 1970 to spring 1971.*

*1988-89 CECC board of directors (standing from left to right) Dale D. Olhausen, Charles C. Crum, Darrel V. Holmquist, Vukoslav E. Aguirre, Gilbert L. Butler, A.S. "Andy" Andrews. (Seated from left to right) Robert E. Emmons, William W. Moore, Richard Weingardt, William Rettberg, Howard Dutzi. Not present: Ken Brotsky, Harold Hollingsworth.*

*(photograph by Ed Bernstein)*

*Past-presidents of CECC attending the April, 1988 installation of officers banquet at DAC, (top row, left to right) Paul Koch 1958, Malcolm Meurer (ACEC past-president 1974), Howard Dutzi 1984, Bill Becker 1970, Larry Muller 1985, George Koonsman 1979, Jim Konkel 1967, Hal Bishop 1980, (middle row, left to right) Ron Blatchley 1977, Fu Hua Chen 1974, Mike Barrett 1982, Jim Stewart 1973, Bob Emmons 1987, (front, left to right) Lee Rice 1975, Jim McFall 1972, Dick Hepworth 1981, Al Anderson 1976, Steve Holt 1986.*

# Beginning Years
## 1955 - 1958

Following World War II, consulting engineers in private practice flourished in the state of Colorado. The most active group was the structural engineers headed by Milo Ketchum. In the late 1940's, a structural engineers association was founded. This was followed in 1950 by the electrical and mechanical engineers forming similar organizations.

Professional Engineers of Colorado (PEC) was established as an association in 1948 and in the period from 1948 through 1960, many consulting engineers served as president. They included Orley Phillips, Alvin Swanson, Ed Thorson and Vernon Konkel. Compared with PEC, and the American Society of Civil Engineers (ASCE), founded in 1908 in Colorado, Consulting Engineers Council of Colorado has a relatively short history. However, their growth of membership and the scope of their activities has had a significant impact on the engineering profession and the progress of the State.

In 1954, the groups of structural engineers, mechanical and electrical engineers consolidated to form Consulting Engineers Association of Colorado. Thirty chartered members were involved, of which 12 were structural, 6 electrical and 12 mechanical. Mr. Kenneth Murray was the association's first president.

The following were the charter members: Rhuel A. Andersen, Frederick G. Anderson, Robert V. Behrent, Hale V. Davis, Jr.,

Rupert J. Fooks, Earl L. Heckman, Clifford Johnson, James J. Johnson, Ib Falk Jorgensen, Milo S. Ketchum, Paul H. Koch, Vernon E. Konkel, James H. Konkel, Harold W. Marshall, Albert E. McKittrick, Kenneth J. Murray, Robert S. Nedell, Orley O. Phillips, Arthur M. Riley, Frederick J. Rink, Joseph W. Sallada, Francis E. Stark, Alvin D. Swanson, Norman H. Todd, John E. Tracy, William W. Trego, Arthur F. Weers, Kenneth R. White, Robert L. Whittlesey, Maurice S. Wilson.

Ken Murray representing Consulting Engineers Association of Colorado attended an organizational meeting of the National Consulting Engineers (CEC/US) in 1955. A constitution and by-laws were adopted. The following were particulars agreed upon at the meeting:

• The initiation fee for members entering the organization shall be $75.00.

• The annual dues shall be $25.00.

• The purpose of the association is to safeguard and improve the consulting engineers' professional status and would raise the standards of the individual consultant's practice.

• The association would serve to augment rather than supplement the work of other societies.

• Eligibility for membership in the association requires the engagement solely in the profession of consulting engineering.

• Engineers who are actually employees of architects, contractors, manufacturers, public utilities or government agencies (and who are, therefore, not truly engaged in independent professional practice) are ineligible for membership.

On July 5, 1956, thirty-nine individuals participated in the first formal meeting of the national Consulting Engineers Council in Tulsa, Oklahoma. Colorado's Kenneth Murray was among these thirty-nine charter members, and he was named as one of the ten initial members of the board of directors. CEC/USA was officially formed on July 6, 1956.

The ten founding state organizations of CEC/USA were:

- Consulting Engineers Association of California
- Chicago Association of Consulting Engineers
- Colorado Association of Consulting Engineers
- Gulf Institute of Consulting Engineers
- Intermountain Institute of Consulting Engineers
- Minnesota Association of Consulting Engineers
- Missouri Association of Consulting Engineers
- New York Association of Consulting Engineers
- Consulting Engineers Association of Upstate New York
- Consulting Engineers Association of Oklahoma

On September 27-29, 1956, the second directors' meeting was held at the Skyline Hotel in Denver. Milo Ketchum, Colorado's second president, presided over this national meeting. Numerous directors' meetings, committee meetings and seminars were held.

Following the directors' meeting in Denver, more states elected to join the national organization. Eventually CEC/USA decided to form a new council under the banner of CEC (Consulting Engineers Council), dropping USA from its name. This name was to continue until 1972 when it adopted the present banner ACEC (American Consulting Engineers Council).

The name change did not affect Colorado until 1960 when the Colorado Association of Consulting Engineers name was dropped and Consulting Engineers Council of Colorado (CECC) was adopted.

Under Colorado's third president, Rhuel Andersen, with Orley Phillips as director to the national CEC, great strides were made. Membership was not limited to structural, electrical and mechanical engineers, and gradually geotechnical and civil engineers joined the association.

In 1958 when Paul Koch took over the presidency, membership had increased to twenty-five companies, and the basic organization was established and has been maintained to this date. In addition to Koch, officers were Robert Behrent, vice-president and Albert McKittrick, secretary-treasurer.

In the early years, association meetings were conducted at various restaurants. Cunningham Restaurant and the American Legion on Speer were favorite choices. It was not until May 1959 that the Denver Athletic Club was selected as a regular meeting place. Since then, for nearly thirty years, the DAC has served as the regular council meeting location.

Orley Phillips recalls that they had no speakers at monthly meetings in the early years; meetings were held to talk about "the business of consulting engineering and about mutual problems."

*Charter members, Koch (center) and Robert Nedell (right), along with Joe Shoemaker and their wives at a CECC banquet in 1958.*

Paul Koch's recollections are that many of the early meetings dealt with whether "we should try to restrict membership by eliminating from consideration those who were borderline engineers, those who cut fees or did shoddy work." He states, "Opinions on the matter were not unanimous; many of our members felt more would be accomplished by getting all — friend and foe alike — in a single room to discuss the problems."

16

Two other topics often discussed were fees and should consultants do engineering on small jobs? Koch says they decided to "stay away from small residences and churches, be available for the job, but not push for it." They established a committee to set fee schedules and guidelines. Because of concern with anti-trust laws, they were careful to "use (these) fee curves for reference only."

During Paul's term one of his main goals was to "get more members." He is unassuming when he reports that membership "almost doubled (during his tenure) to over sixty members." Liability insurance costs were not a major factor. He paid $300 annually for $100,000 in coverage with Lloyds of London. Later in 1979, when he employed 15 people, his premium increased to $3,500 per year for $1 million in coverage with a $5,000 deductible.

*The 1958 national meeting was held in Dallas. The meeting brought delegates from 20 member associations representing 1000 firms. Colorado was represented by president Paul Koch and national director Orley Phillips. (Koch is 5th from left; Phillips is 8th from left.)*

# ORLEY O. PHILLIPS

ORLEY PHILLIPS was the first member from the Colorado Association of Consulting Engineers elected as a national officer, vice-president of CEC/USA (ACEC), 1959-60. He was born in Victor, Colorado, November 24, 1903 and received his Bachelor of Science degree in civil engineering from the University of Colorado in 1926.

His long and varied career has included engineering projects throughout the United States; Colombia, Venezuela, Peru, Brazil, and Argentina in South America; Pakistan; Mexico; Spain; Newfoundland and Canada. He first worked for the old Dominion Copper Company in Arizona then with the American Smelting and Refining Company in Texas, Mexico *and Newfoundland. From 1931-1933 he was an engineer for the Buffalo Exploration and Mining Company in USA, Mexico and Canada. He returned to Colorado in the mid-1930's as the assistant city engineer for the city of Colorado Springs. Subsequently he became the chief structural engineer for the Holly Sugar Company where he was in charge of structural design for a large sugar beet factory.*

*In 1937 he became a partner in the firm of Sanford and Phillips, consulting engineers in Colorado Springs, designing and supervising construction of structural steel and reinforced concrete structures, including highway bridges, viaducts, industrial buildings, hotels, schools, an all-steel stadium, waterworks, dams, pipe lines, and sewage systems.*

*Phillips joined the firm of R.J. Tipton and Associates in 1944, after a brief period with Stearns-Roger as engineer-in-charge of expediting construction of an aviation gasoline plant. From 1949-57 he was vice-president for R.J. Tipton and Associates in charge of operations in the USA. The firm of Phillips-Carter-Osborn consulting engineers was then formed and he served as its president.*

*From 1966-73 he was president of Phillips Carter Reister and Associates, architects and engineers. In 1973 he became a vice-president in Daniel Mann Johnson and Mendenhall and the executive vice-president of the Denver office, DMJM-Phillips Reister Haley, Inc. Significant Colorado water supply projects he worked on include Roberts Tunnel completed in 1962 at a cost of $51 million, Vasquez Tunnel, and Foothills Tunnel and Conduit for the Denver Board of Water Commissioners. His favorite project was the design of Mile High Stadium from the first baseball field plus all expansions including the award winning movable east stands.*

*Outside engineering societies, Phillips has served in many advisory capacities and has donated freely of his time. His civic activities included serving as chairman of the Denver Building Code Advisory Committee, chairman of the University of Colorado Engineering Development Foundation, and chairman of the Colorado State Board of Registration for Professional Engineers. He was a member in the Denver Chamber of Commerce and Downtown Denver Improvement Association.*

*He has received the ASCE Edmund Friedman Professional Recognition award, the 1970 Distinguished Engineering Alumnus in private practice award from the University of Colorado, the ACEC Award of Merit, the Colorado Engineering Council Honor award and the Gold Medal award. He was the inspiration for and the first recipient of the CECC Orley Phillips (outstanding engineer of the year) award. He is an honor member of Chi Epsilon fraternity.*

*Orley and his wife continue to make Colorado their home. They live in Denver and have one son.*

*Jim Konkel (left), CECC president-elect, and Gene Waggoner, ACEC president, discussing council activities and an April 1967 memo from Hak Kadish. Waggoner was the first Colorado member to be elected to the lofty position of president of the national association.  He also served as ACEC vice president 1962-63.*

# Growing Period

## 1958 - 1964

Under the leadership of CECC presidents, Alvin Swanson, John Tracy, Eugene Waggoner, E.H. Carter and Vernon Konkel, the status of the Council was firmly established among the engineering community in Colorado. CECC had reached the stature of ASCE and PEC, the other leading engineering organizations in the state.

Orley Phillips, first as Colorado's national director then as an ACEC vice president, likewise was keeping CEC/Colorado visible as a force on the national scene. When Phillips was elected vice president of the national association in 1959, it marked the first time a Colorado consulting engineer held an elected office in ACEC.

Taking over as Colorado's national director when Phillips assumed his duties as vice president of ACEC, was Vern Konkel, CECC's second national director. Konkel was energetic and his leadership skills influenced national issues to the benefit of Colorado. He was a U.S. delegate from the Council to the International Federation of Consulting Engineers in Helsinki, London and Paris. He served as Colorado's national director for three years until his election as president of CEC/Colorado.

Charles Meurer became the local association's third national director in 1963. Gene Waggoner was elected as vice president of the national CEC organization from 1962 to 1963, the second Colorado member to so serve.

The need for consulting engineering firms increased along with Colorado's post-war population growth which continued in the 1950's at a rate of 2.8 percent per year; and in the 1960's at 1.9 percent.

A significant public works project on which consulting engineers such as Orley Phillips, Al Ryan and Herbert Crocker played an important role was the Valley Highway. This major metro-Denver highway was completed in 1958, the same year the $140 million United States Air Force Academy was completed in Colorado Springs.

In June 1956, President Eisenhower authorized $33.5 billion for a nationwide network of highways linking most major cities in the USA. This federal Interstate system was to be called the "greatest public works program in history." Federal money for design and construction of I-25 and I-70 was infused into the Colorado economy.

The Soviet Union shocked Americans by putting the first man in space in 1961. USA President Kennedy officially commited America to put a man on the moon "before this decade is out."

Then on November 22, 1963, John F. Kennedy was shot and killed. Colorado, as well as the rest of the nation and the world, was stunned. Possibly in his memory, we gained a stronger national fervor to harness our scientific and technical capabilities to put an American on the moon in the 1960's. Engineers across the nation played a major role in this effort.

The independent practice of consulting engineering was coming into its own.

Under the presidency of Vern Konkel, the affairs of the Colorado Council became so massive that a decision was made to hire an executive director. Harvey Kadish was appointed to this position in August, 1963. Since the public relations activities are inseparable from the operations of the association, the post of CECC executive director was created as a joint PR and management function.

Kadish, affectionately referred to as "Hak", was the author of many articles in engineering and technical publications. His achievements and service to the consulting engineering profession are legendary.

According to Paul Koch, Hak was very "effective with state legislators" and a big influence in getting CECC scholarships started. Also, he began the concept of "having speakers at regular monthly meetings." Several council meetings were held at the Press Club on Glenarm in downtown Denver because Hak was a member of the Press Club.

The first truly national annual membership convention meeting of CEC/USA was organized and managed by CEC/Colorado in May 1964. The meeting at Denver's Brown Palace Hotel drew more than 550 members, wives, speakers and guests, with 42 of the 50 states and 41 of the 43 member organizations in attendance. Vernon Konkel, then president of CEC/C and vice president of CEC/USA presided. Even today, it is remembered as one of CEC's most successful events.

The growing period came to an end. Colorado's influence was being felt at the national level.

---

*Consulting engineers would soon be very involved in two major, natural disasters about to hit the state; the June 1965 South Platte River flood and the 1975 Big Thompson flash flood. Both cause major destruction washing out buildings, bridges, roads and other improvements. Engineers would play a major role in the massive repair and replacement efforts.*

*Part of the Colorado delegation at the national ACEC convention 1972. (From left to right) Harvey Kadish, Jim McFall (CECC president), Fu Hua Chen (CECC national director) and Andy Pfeiffenberger. It was the year the official name for the national Consulting Engineers Council was changed to its present name, American Consulting Engineers Council (ACEC). 1972 also marked a milestone in the federal government's selection process for hiring consulting engineers in private practice. The Brooks Bill, sponsored by Texas Congressman Brooks, established the use of qualification-based selection procedures rather than bidding be used when hiring professionals such as engineers and architects. Harvey "Hak" Kadish served as CECC executive director from 1963 until his death in 1973.*

# Leaders Nationally

## 1964 - 1976

In 1964 when John Bunts took over as president, coordination between consulting engineers (CECC) and professional engineers (PEC, NSPE) became a high priority. An example of a successful joint program dealt with the issue of "Employment Practices"; to study fees and salaries in governmental organizations. Speakers were John Stamm, for engineers in government, and James Smith representing NSPE and their PEPP section.

Under president Bunts, the public relations committee chaired by Kenneth Wright became active and branched over into legislative concerns. Colorado participated for the first time in a visit to Washington, D.C. to discuss legislative matters. The practice of extending invitations to state legislators to attend a CECC general meeting was initiated. This practice continues to this day.

A speakers bureau was established for the purpose of educating the student on the role of consultants. Fu Hua Chen was appointed by Bunts as chairman of this committee.

Charles S. Meurer followed John Bunts as president. Charlie was a partner in the powerful engineering firm, Meurer, Serafini & Meurer. Malcolm Meurer, another partner in the firm, was editor of the association's newsletter which was called the "Bulletin."

Meurer, in his inaugural speech as he began his term of office, listed three aims of his administration: "to combat attempts of unions to organize employees of consulting engineering firms, to increase the business of consulting engineers, and to establish better relations with colleges of engineering."

In addition, he emphasized, "The continuing fight for engineers in private practice is to get a greater share in governmental work and to ask the state legislature to pass a statute of limitations bill."

The Consulting Engineers Council of Colorado now had a membership strength of 66 firms. Individually, 86 consulting engineers displayed the membership certificate. Member firms were engaged in practically every discipline of engineering.

William Clevenger, who later became the 22nd vice president of ACEC, was president of Colorado CEC in 1966. A big issue continued to be the problem with unions. The union tried to organize the field personnel of Phillips-Carter-Osborn and Henningson, Durham and Richardson. A union election was to be held in June 1966. Bill Clevenger pitched in immediately to help the two member firms and went on record to pay the cost of an experienced labor lawyer. Both firms had many and long negotiating sessions with operating engineers (Local No. 9 AFL-CIO).

It was concluded that unionization attempts might be curtailed in the future if consulting firms would maintain good communications with their employees, pay fair wages and provide fair benefits. That has been the case.

*Bill Clevenger, relaxing away from the cares of office and CECC.*

Since Clevenger was a geotechnical engineer, it is understandable that he would establish a "professional consulting soils engineers" ad hoc committee, where common issues were discussed. Twenty-three years later, geotechnical engineers as a group still adhere to the general principles discussed and agreed upon at that time.

In 1966, CEC/Colorado was well represented on the national scene. Both Vernon Konkel and Eugene Waggoner were national officers; Gene was ACEC president-elect and Vern was vice president. The Colorado association's national director was Ken Wright.

From 1963 until 1975, CEC/Colorado had more than its share of officers serving the national association (ACEC); two presidents, Waggoner and Malcolm Meurer; three vice presidents, Clevenger, Waggoner and Konkel, and a secretary-treasurer, Meurer. Leadership by Coloradoans was definitely impacting national issues!

CEC/Colorado initiated its annual engineering excellence awards program in 1967, to recognize engineering achievements by its members that demonstrate the highest degree of ingenuity and skill; contributing to technical, economical and social advancement. The first recipient of these prestigious awards was the Ken R. White Company for the engineering of the Ideal Cement Plant in Seattle, Washington.

James Konkel, who was president in 1967, received the 1984 Orley Phillips Award for his "many outstanding contributions to the consulting engineering profession and the Council (CECC)."

While William Hawes was president in 1968, CECC issued its "Code of Ethics." The code was adopted from national ACEC documents. The contents of the code came under much scrutiny in later years, especially in 1975 when NSPE was found in violation of the Sherman Antitrust Law. A new guideline was adopted in 1980.

*President-elect Ken Wright (right) explains his firm's 1969 award-winning project to Richard Lamm, a young Colorado state legislator who would one day become governor.*

On September 29, 1968, CECC placed a full page ad in the "Empire" magazine of The Denver Post explaining the association. Covered were items such as "how to select an engineer, using a consultant, and when to seek service of a consultant." The following extract is illustrative of what was in the ad:

*CEC/Colorado, The Consultants'*
*Organization*
*The Consulting Engineers Council of Colorado (CEC/Colorado) is an organization of more than 100 registered professional engineers, each a principal of a consulting engineering firm with a practice in the centennial state.*

*The more than 77 firms, represented by CEC/Colorado, are engaged solely in the private practice of engineering. They have no commercial affiliations with manufacturers, suppliers or contractors; thus, they are able to render unbiased engineering decisions both for design and choice of material.*

On July 20, 1969, shortly after Ken Wright took office as CECC president, two USA astronauts walked on the moon. Neil Armstrong, educated as an engineer and from Colorado, was the first man to step on the moon. Armstrong commented, "That's one small step for a man, one giant leap for mankind." Many historians say this was the greatest scientific achievement in history.

A move was made in 1969 in Washington, D.C. to consolidate the engineering organizations. It was proposed that Consulting Engineers Council, the American Institute of Consulting Engineers and the Professional Engineers in Private Practice (PEPP) be consolidated. CECC, represented by president Wright, went on record as opposing the merger. The talks of merger however were not over yet.

Around this same time, the CECC Code of Ethics was challenged. Fu Hua Chen, chairman of the Ethical Practice committee, presented a lengthy report outlining apparent violations of the code in regard to the Metro-Denver sewage dispute. Hearings were held and concerns discussed.

Eventually CECC members voted on a revision to the Constitution and the By-Laws with a completely new section, "Guide for Professional Conduct." Formal

"Articles of Incorporation of Consulting Engineers Council of Colorado" were filed with the state of Colorado, August 3, 1970, under signature of Ralph W. Becker, president, and Fu Hua Chen, secretary-treasurer. In addition to Becker and Chen, the CECC Board of Directors listed on the document are: Michael Barrett, David Fleming, C. Kenneth Kolstad, George Koonsman, Malcolm Meurer, James McFall, Orley Phillips, James Stewart, Alvin Swanson, George Williams, and Kenneth Wright.

Incorporation details were handled by Joe Shoemaker who was retained by the CECC Board as legal counsel. Joe was the only person in the Colorado state legislature at that time with an engineering background.

The USA now had 203 million people.

The population of Colorado in 1970 was over 2.2 million, a 25 percent increase over 1960. The state was experiencing a lot of growth. Currigan Hall, Denver's new convention center, started attracting national conventions.

The first CECC organization chart was established by President David Fleming in 1971. The basic organization has been maintained since.

1971 was "Olympic talk" year for Colorado. The world was witnessing larger and larger participation in the Olympics and Denver, Colorado, was designated as the site of the 1976 Winter Olympics. There was a lot of "excitement in the air" and CECC became part of it. Ken Wright was appointed to the Olympics Planning Commission. A special and dynamic committee was established and chaired by Robert Barber. The goal of the committee was to provide assistance to the Denver organizing committee, for the benefit of Colorado. Unfortunately, before anything really got under way, the entire Olympics effort was cancelled by Governor Dick Lamm.

The ACEC board of directors at its semi-annual meeting in October 1971 again studied merger; a proposal for the national consolidation of the consulting engineering profession was reviewed. The consolidation would combine ACEC, NSPE-PEPP and AICE. A poll of CEC/Colorado members overwhelmingly rejected the consolidation concept. Discussions on a merger were finally put to rest.

Bidding fees became a hot issue in early 1972. President Fleming was a strong spokesman against the concept stating, "Members of CEC/Colorado voluntarily adhere to a code of ethics upholding the principle that the selection of a consulting engineer should be on the basis of qualifications, including training, skill, experience, personnel, work loads, and availability."

The Code of Ethics, according to Fleming, "outlines the engineer's responsibility to safeguard health, safety and public welfare, and professional consulting engineers generally do not enter into competitive bidding, since it is not in the public welfare."

Fleming added, "To draw an analogy, I just can't imagine a person needing brain surgery trying to get a low bidder to perform the operation. Instead, one would attempt to get the most competent surgeon he could find, and one thoroughly capable by education, experience and skill. Engineers, too, are professionals, and it is in the client's interest to select his engineer on the basis of technical competence and experience."

A newly-published CEC booklet was widely distributed outlining the recommended process for the selection of a consulting engineer, stressing selection on capabilities and qualifications before negotiating a fee arrangement. It further stressed someone using professional services was a "client," and not a "customer," and a professional was "retained," not "hired."

A most important milestone for Consulting Engineers Council was the signing into law of the A-E Selection Procedures Bill introduced by Texas Congressman Brooks. Consulting engineers would be selected for federal projects based on qualifications; not by bidding professional services. In 1972 representatives joined officials of other engineering and architectural societies in the oval office of the White House receiving first copies of the signed act. The now well-known "Brooks Bill" was a landmark victory for consultants and other professionals.

Because of president James McFall's efforts, Colorado senators Pete Dominick and Gordon Allott solidly supported Colorado consultants on this bill.

Other notable events occurring in 1972 under president McFall were; negotiating with the Environmental Protection Agency (EPA) for the design of $5 billion worth of wastewater treatment plants, and addressing the dam safety inspection guidelines.

By now CECC was realizing that consulting engineers must be more involved in governmental and legislative activities. Ron Blatchley, personally on behalf of the Council, took extreme interest in the legislative arena and attended many hearings concerning issues affecting engineers. His efforts paved the way for CECC to play an important role in legislative affairs for the years to come.

With the support of Senate majority leader Joe Schieffelin, legislation important to consultants was passed and was in the making under 1973 CECC President, James Stewart. Colorado SB-35 on land use regulations was enacted in 1972. This encouraged the adoption of the "Land Use Regulations" by the Colorado Land Use Commission in 1976. The entire process was four years in the making, with active participation from CECC.

By 1973, the CEC/Colorado budget was adequate to allow for a full-time headquarters office. Harvey Kadish dissolved his Colorado public service office and devoted his entire effort as the executive director of CEC/Colorado. Sadly, Hak died in October 1973.

One of the highlights of the annual convention at New Orleans in 1973 was the naming of 15 winners in the annual Engineering Excellence Awards program. The top award, the grand conceptor, went to Ketchum-Konkel-Barrett-Nickel-Austin, of Denver, Colorado, for its structural design of TWA's thin shell, high bay aircraft hanger, located in Kansas City.

Early in 1974, the CECC ad hoc selection committee recommended the engagement of Edward Thorson to succeed Kadish as executive director. Ed had a long history of accomplishments in engineering organizations, including being president of Professional Engineers of Colorado in 1958.

It appears that for the years to come, legislative issues will continue to be a main concern for CECC and its members.

Increasing premiums for errors and omissions insurance, frivolous lawsuits, and skyrocketing legal costs were now demanding tort reform.

It was decided in the board of directors meeting on January 7, 1975, to retain "Pat" Patrick as a "legislative observer" to keep the Council informed of developments at the statehouse that might be of interest to CECC members. Pat was so successful in his activities that his role with CECC steadily increased. Mostly because of his efforts, the Council is now recognized and respected in the legislative arena.

The most important piece of legislation sponsored by CECC during President Fu Hua Chen's tenure was HB 1210. Consulting engineers believed that a starting point in trying to curtail the trend of unchecked liability suits would be legislation authorizing the award of court costs and reasonable defense attorney's fees, to any engineers found innocent in cases of a frivolous suit. HB 1210, commonly referred to as the "Frivolous Suit Bill," was finally enacted in 1977. The bill, as seems normal with much legislation, is not exactly what was intended, but is considered to be a good starting point. The bill was amended in 1987 to enlarge its scope to cover other professionals.

Representative seminars sponsored by CECC while Fu Hua was president included two very successful ones; "The Consulting Engineers and the Environment" symposium held April 5, 1974, and chaired by Joe Meheen, and "Engineered Development and the Law" symposium held October 25, 1974.

Colorado's Malcolm R. Meurer was elected as the 19th president of National Consulting Engineers Council which made him the second president of the newly consolidated organization now officially called the American Consulting Engineers Council (ACEC).

Margaret Bates Ellison was retained by CECC as a consultant for legal matters. Her first assignment was the drafting of a proposed revision to the state's professional engineer's registration law. The exclusion of certain engineers working for governmental agencies from the requirement to be registered has been of great concern to the Colorado Board of Registration as well as Professional Engineers of Colorado. It has been a hot issue for many years and an end to this issue has yet to come.

---

*Several CECC members played important roles serving on the Colorado State Board of Registration for Professional Engineers and Land Surveyors:*
*1963-1974 Orley Phillips **
*1974-1978 Arthur Krill*
*1978-1981 Fu Hua Chen*
*1975-1981 Kenneth Wright**
*1980-1989 George Sellards**
*(*served as chairman several years)*

---

During 1974, because of the efforts of Bill Hawes, close relations were developed with Associated General Contractors and Colorado Contractors Association. Pat Patrick and Pierre Dubois, legislative consultant for the contractors, worked closely on bills such as "Damage Resulting from Delays in Contract Performance Resulting from Injunctions Wrongfully

Obtained" and, of course, the "Frivolous Suit Bill."

Six Coloradoans were named as ACEC Fellows in 1975. Elected to the honor were Malcolm Meurer, Rhuel Andersen, Fu Hua Chen, David Fleming, Orley Phillips and Alvin Swanson. To be eligible for admission to the ranks of fellows, an individual must have contributed notably to the advancement of consulting engineering in one of eight categories: administrative leadership, design, science, construction, education, literature, service to the profession or by public service.

With Leonard Rice as president in 1975, CECC was actively involved with environmental issues, especially EPA guidelines regarding grant regulations and procedures.

Colorado SB 239, referred to as "our" bill, delineated supervisory responsibilities of engineers on construction projects, and was enacted as law.

At the national level, the United States District Court decided that the NSPE Code of Ethics' prohibition of competitive price proposals for professional services was "in every respect a classic example of price fixing in violation of the Sherman Antitrust Act." This was a blow to professionalism. CECC claimed that, "Imposing such a system upon the selection of professionals can result only in reducing the quality of services, and eventual damage to those whom the courts and the Justice Department think they are protecting." ACEC pledged to continue supporting the NSPE effort to negate the antitrust action and the court decision.

In his eloquent "state of the Council" address April 22, 1976, Lee Rice pointed out, "after 21 years of service, CECC has met many challenges to our profession and has contributed much to our state and country. In 1975 member firms employed directly over 2000 professionals and technical people with an aggregate payroll exceeding $30 million. This payroll generated some $120 million in goods and services for the economy of Colorado. Individual members were responsible for the planning, design and supervision of construction of more than $500 million worth of public and private facilities that support the standard of living that we enjoy in this great state."

After delineating the successes and disappointments of his time as president, Lee suggested the issues CECC would face in the coming year. He then concluded, "At a time when it is fashionable to criticize our institutions and traditions, let me say that I am proud to be a United States citizen living under the greatest form of government yet devised by man. I am proud to be a businessman trying to make the free enterprise system work, and I am proud to be a consulting engineer contributing to the improvement of our environment and the quality of our life."

And with that, the Council and its members prepared themselves for a construction boom, the likes of which the state of Colorado had never seen before.

*Governor Dick Lamm signs National Engineers Proclamation for February 1976; Witnessing (left to right) are Lee Rice, CECC president, Gilbert Rindahl, David Day and Dr. Albert Knott.*

# Boom Years
## 1976 - 1982

The "face of downtown" Denver was virtually changed in a half dozen years, between the middle seventies and 1982. Construction cranes became so commonplace they were jokingly called the Colorado "state bird." When the 1970's were ushered in, Denver had fewer than a dozen buildings over 300 feet tall, with Brooks Towers being the tallest at 420 feet. By 1982 ARCO tower at 608 feet was the tallest structure. There were more than 25 high-rises over 300 feet completed and in use, with several more under construction. CECC firms were engineers or consultants on many of the projects.

What at the time did not seem to concern many was that over one-third of all downtown Denver office space was occupied by energy companies. Energy prices were sky high. Too high to last.

Albert Anderson, a native Denverite, born, raised and educated in the mile-high city, was CECC's president in 1976. A structural engineer, Anderson was only the fourth person to date in that discipline elected to lead the association.

Recently, when asked to reflect about his term as president and the profession, Al commented, "Our problems are insurance costs, RFP responses that are dollar bids only and the professional image of consultants." He then allowed that those issues have been around "for 30 years and probably always will be."

Structural engineers were busy those days designing the many new structures appearing almost overnight throughout Colorado. Nearly half of the CECC board of directors were structural engineers: Anderson, Ed Bierbach, Donavon Nickel, Bill Hawes and Joe Meheen.

*Leslie Botham (left) of Leonard Rice Consulting Water Engineers receiving the firm's Excellence Award for their project, "Evaluation Procedure for Urban Drainage at CSU" 1976. Congratulating Botham is H.O. (Mac) McElyea, chairman of the judges for the Engineering Excellence Awards program.*

Ron Blatchley, who followed Anderson as president, recollected about his term with: "The most enjoyable part of serving CECC membership has been the legislative involvement. This started long before I was part of CECC and before becoming president. After joining CECC, I was able to be on the leading edge in forming CECC's legislative policy and organization."

The association's political action arm, CEPAC, was officially incorporated in 1978 under the direction of president Blatchley. Ron was its president, Pat Patrick its vice president and Ed Thorson was secretary-treasurer. CEPAC is alive and well today working in close association with PEC's PEPAC. The two PAC's continue to increase their activities.

Two Colorado engineering achievements attracted national attention in 1978; the "movable east stands" of Denver's Mile High Stadium and Lamar's bioconversion facility.

The "movable east stands" designed by DMJM-Phillips, Reister is the largest structure ever to be moved by the "fluid film" water-bearing system. The project was the recipient of three national engineering awards for outstanding achievement. Governor Lamm stated in a proclamation, "It is one of the unique engineering accomplishments in the world today."

The American Consulting Engineers Council's top engineering award, the

Grand Conceptor Award, was presented to CH2M Hill for their design of a bioconversion facility in Lamar, Colorado. The "biogasification" process is engineered to convert manure into methane gas and cattlefeed. The actual facility was never constructed, however. Because of the eventual turnaround in energy prices it became no longer economically feasible.

Upon taking office as president in 1978, Donavon Nickel commented, "The Consulting Engineers Council of Colorado is dedicated to serving its members in the area of business organization, allowing them to do a better job of managing their business and consequently being able to provide a better service to their clients."

Towards the end of Nickel's term, the association had grown to 168 member firms with 248 principals as individual members. Member firms employed 2700 people.

1979 was the year of the "energy crisis." Service stations were lined up with cars from sunrise to sunset. Gasoline was rationed and prices were outrageous. Skyrocketing fuel prices forced the government belatedly to consider the development of oil shale. CECC rose to the challenge of energy conservation by initiating several programs, including energy audits, solar energy use, oil shale development and gasohol development. The demand increased for the services of mechanical and electrical engineers as well as geologists and mining engineers.

Following the guidelines of the federal "Brooks Bill," CECC introduced the "Mini-Brooks" Bill (HB 1432). The bill was passed in 1979. It required that all state agencies select and retain engineers, architects, landscape architects and surveyors based on qualifications.

"The statute assures Colorado of utilization of the best qualified consultants in future works," declared CECC's 1979 president, George Koonsman.

During his year as president, Koonsman fought hard with the EPA on the delays being imposed on the Foothills Water

Treatment Plant. The project was eventually completed but at a considerable increase in cost over original estimates because of these delays.

Koonsman brought to everyone's attention the tremendous contribution consulting engineers were making to the state's economy. George emphasized that, "Considerable engineering is being exported from Colorado to other states as well as overseas. This results in a positive economic impact and increase in the number of jobs within the state."

Of particular interest in the global marketplace in 1979, was the potential for providing engineering services to mainland China. Colorado's Fu Hua Chen, with a contingent of twenty consulting engineers from all over the country went to China to reconnoiter and lay the ground for working with that country.

Harold Bishop took over as president in 1980 as the state's population reached 2.9 million people. Hal said, "My recollec-

tion was that bidding, government competition and increasing membership all seemed to play an important part in my year as president." He added, "I believe interchange of thoughts is another one of the most valuable things gained from CECC. It is interesting that we can sit down and talk business problems with our competitors and maintain a personal relationship with them."

*Edward W. Thorson, Executive Director of CECC 1974-1981.*

In June, shortly after Bishop became president, CECC's executive director, Ed Thorson, was taken to Swedish Medical and was not able to resume his duties.

40

The Council hired Gail Bays, past managing director of administrative services at ASCE, in September.

CECC, with president Bishop presiding, hosted candidates for the U.S. Senate and U.S. Congress at the October 28, 1980, general meeting.  Attending were Patricia Schroeder, Gary Hart, Tim Wirth and Kenneth Kramer, key political figures in Colorado as well as nationally.

The presidency of CECC was transferred to Richard Hepworth in 1981.  Sondra Smith became the fourth executive director of the association, replacing Gail Bays who had resigned.  Sondra was previously on the management staff of the Lakewood Country Club.

*Ellen & Larry Muller with Deryl Gingery (seated) at the April, 1981 general meeting at the DAC.*

It was a great loss to Colorado's engineering community as well as the Council when Ed Thorson passed away in 1982, concluding his long service to engineering societies in the state.

The same year, Fu Hua Chen was elected vice president of ACEC, the fourth Coloradoan to be elected to the position.  Al Anderson was CECC's national director in 1981 and Jim Konkel assumed the position in 1982-83.

Chen received the prestigious CECC Orley Phillips Award in 1982 for his many contributions to the profession and the Council.

There was another attempt at the national level to consolidate the programs of American Consulting Engineers Council and National Society of Professional Engineers in 1982.  It was tentatively agreed that consolidation of ACEC and NSPE at the national level would not mandate similar consolidation of the state or local level.  The consolidation at the national level failed to get enough support and the merger attempt failed again.

Hepworth's accomplishments at the local level were outlined by Dick, "I think the best thing we accomplished was to hire Sandy.  Other things were the continued improvement of our financial status budgeting for capital improvements establishing liaison with the State Engineer and establishing a western slope chapter. Probably the least considered action I took was to formulate the policy memo regarding competition from the Universities." Hepworth was and continues to be a strong proponent of an active CECC "Speaker's Bureau."

Membership in the association was now 237 firms employing 3700 people.

In 1982 the state population exceeded 3 million people for the first time, then suddenly growth slowed drastically.

The boom years were over and the Colorado economy started its tailspin.

The Lawn Lake Dam above Estes Park collapsed.

The "blizzard of 82", the worst snowstorm in 69 years, buried the state under 2 to 3 feet of snow on Christmas day.

*The 1979 CECC Engineering Excellence Awards Banquet was held at the 26 Club of the First National Bank Building on April 25th; one of the few times the annual event was not at the Denver Athletic Club. Back in those days the awards presentations and the installation of officers occurred at the same banquet.*

*L to R, Dick Sittner, Dave Flemming, Fu Hua Chen*

*L to R, Stephen Holt and John L. Haley*

*L to R, Don & Charla Berry, Evie Weingardt*

# Early Eighties
## 1982 - 1986

The failure of the Lawn Lake Dam caused a great deal of turmoil and triggered a series of actions involving the State Engineer, legislators, engineers and the insurance industry. Key issues addressed were: the State Engineer's responsibilities, insurance coverages and the role of private consultants in dam safety inspections.

Colorado Water Congress conducted meetings under an ad hoc "Dam and Reservoir Safety" Committee and CECC members, Ron Halley, Richard Hepworth, Ken Wright, Ron Blatchley, Fu Hua Chen and Pat Patrick attended all meetings. They contributed to a series of important legislation enacted in 1983 on the dam safety issues.

The gross annual budget of the Council by 1983 had grown to $143,000, a 200 percent increase from 1973, ten years earlier.

The public relations committee, through the leadership of several energetic chairmen, became — and stayed — very aggressive in the early 1980's. At the direction of Dick Hepworth, they retained Evie Weingardt of Jacqueline Enterprises as a consultant and produced the first in a long series of CECC calendars featuring "historic Colorado projects." The P.R. committee also placed ads depicting "the glories of consulting engineers" and worked with local media on news releases and other features. The Daily Journal's coverage of CECC's Engineering Excellence Awards program and banquet was expanded.

The CECC By-Laws were revised in 1983. Colorado was geographically divided into four districts with Denver being District 1. The southern, northern and western slope districts had become very active locally which eased the concern that consultants outside of the Denver Metropolitan area were not able to attend monthly meetings held in Denver. Ed Carpenter was a prime mover in getting the western slope active. Elmer Claycomb and Roger Hocking were officers.

Membership in CEC Colorado increased steadily. The voting power of the Colorado association — an index number of 935 — was third nationally in 1982. Only California and Texas were larger. Considering that the population of Colorado is far less than Texas and California, CECC took pride in its achievement. Plans for a "Bigger than Texas" membership campaign were developed. The campaign headed by President Mike Barrett created a great commotion at the CECC annual meeting in San Antonio.

Commented Barrett, "We had a number of significant accomplishments . . . (such as) stirring up some excitement about membership, including competition with the state of Texas for attendance at national meetings and total growth of state membership."

Colorado's membership eventually increased to 314 firms, employing 4,834 persons and a national index number of 1,243. But it never surpassed California or Texas. Their state population growths were too great.

James Hastings, who succeeded Mike as president, kept alive the campaign to increase membership. Jim also pushed for better cooperation between engineering societies in the state. Given President Hasting's directive, the long range planning committee recommended, "Every effort should be made to cooperate with other engineering societies in matters of mutual interest. The annual Engineers Dinner, Engineers Week and <u>Colorado Engineering</u> (magazine) are examples of such activities."

James (Jim) Speer, executive director of the Professional Engineers of Colorado, was killed in an auto accident in 1984 while vacationing in Ireland. He had done much in the efforts for cooperation between CECC and PEC. The two engineering groups, though having different agendas and membership profiles, continue to work together with their political PAC's and their legislative issues.

During the past few years, the relationship between CECC and the state legislature has steadily increased. At the Capitol, many legislators now respect the input of consultants. Engineer's joint political action committee efforts have been a success. "Meet Your Legislator Night" has become an annual event, jointly sponsored by CECC, PEC and the universities.

Arnie Windman, president-elect of ACEC, Howard Dutzi, 1984 president of CECC, and a handful of association "veterans" called on several large non-CECC engineering firms to join the council. Howard proudly comments, "These visitations were highly successful. All of these firms eventually joined CECC, but only one, as I recall, joined during my tenure as president."

Dutzi, a Colorado Springs structural engineer, is the 1988-90 national director for the Colorado association. In recent correspondence, he contemplated, "Over the years I have received much input from our various members and it seems to me that they are basically interested in how they can better manage their firms. It therefore seems to me that (CECC) programs which affect our businesses have much appeal."

*Howard Dutzi with well-wishers at his installation as president (left to right, Steve Holt, Larry Muller, Dutzi, Art Greengard).*

Ed Carpenter, past chairman of the CECC Western Slope chapter, was elected to the Colorado House of Representatives on the Republican ticket.

Many considered 1983/84 the peak of Colorado booming economy. The declining years were to follow.

---

*In October 1985, past-president Howard Dutzi and CECC hosted the ACEC Annual Meeting, held at the Broadmoor Hotel in Colorado Springs. The meeting was one of the best ACEC annual meetings.*

*Consulting engineers in Colorado identified the major projects they worked on in 1983/84 as follows:*
*Commercial/office projects 29%*
*Housing 19%*
*Water supply/resources 17%*
*Transportation 6%*
*Industrial facilities 5%*
*Public buildings 4%*
*Wastewater facilities 3%*
*Educational facilities 2%*

*CECC moved its office from 1111 South Colorado Boulevard to 940 Wadsworth Boulevard, in July 1985.*

*Governor Lamm displays Engineers Week Proclamation he has just signed, 1986 (left to right, John Kreiling, Larry Muller, David Sveum, Fred Zook).*

46

Initiation of legislative tort reform and the reorganization of the CECC board were significant accomplishments of the Larry Muller presidency. The reorganization was done, according to Larry to, "require greater involvement of the directors in the operation of the Council. Steve (Holt) carried this through."

Skyrocketing liability insurance premiums and lack of available insurance were the catalyst for tort reform. Muller, Pat Patrick the association's political consultant (lobbyist), and legislative committee chairman, Fu Hua Chen, along with several others including Richard Weingardt, Andy Andrews, Steve Holt and Sandy Donnel spent long hours at the State Capitol, at hearings and with state legislators discussing tort reform.

Bill Bredar, chairman of the ACEC professional liability committee, also played an important role in the national tort reform legislation. The Colorado tort reform effort has been hailed by many states in the Union as model legislation.

In response to the statewide tort reform movement, CECC issued a position paper on the "Professional Liability Crisis in Colorado" in October 1985, as follows:

The Consulting Engineers Council of Colorado and its more than 300 member consulting engineering firms in private practice, who employ more than 4,400 people in Colorado take the position that the legislature shall:

*1. Enact laws to limit compensation for claims regarding damages for liability awards to actual damages incurred. Punitive damages to be awarded only in case of proven fraud.*

*2. Limit contingency fee for suits with respect to professional liability claims.*

*3. Introduce statute requiring an attorney to file certificate of consultation with design professional of same discipline to substantiate that the suit does have merit.*

The American Consulting Engineers Council was pursuing similar legislation throughout the United States. ACEC was now composed of 4,434 member firms employing 126,700 persons.

Trial lawyers and insurance industry officials were blaming each other for the insurance crisis, unreasonable lawsuit judgments and the lack of available liability insurance, at any cost, for businesses.

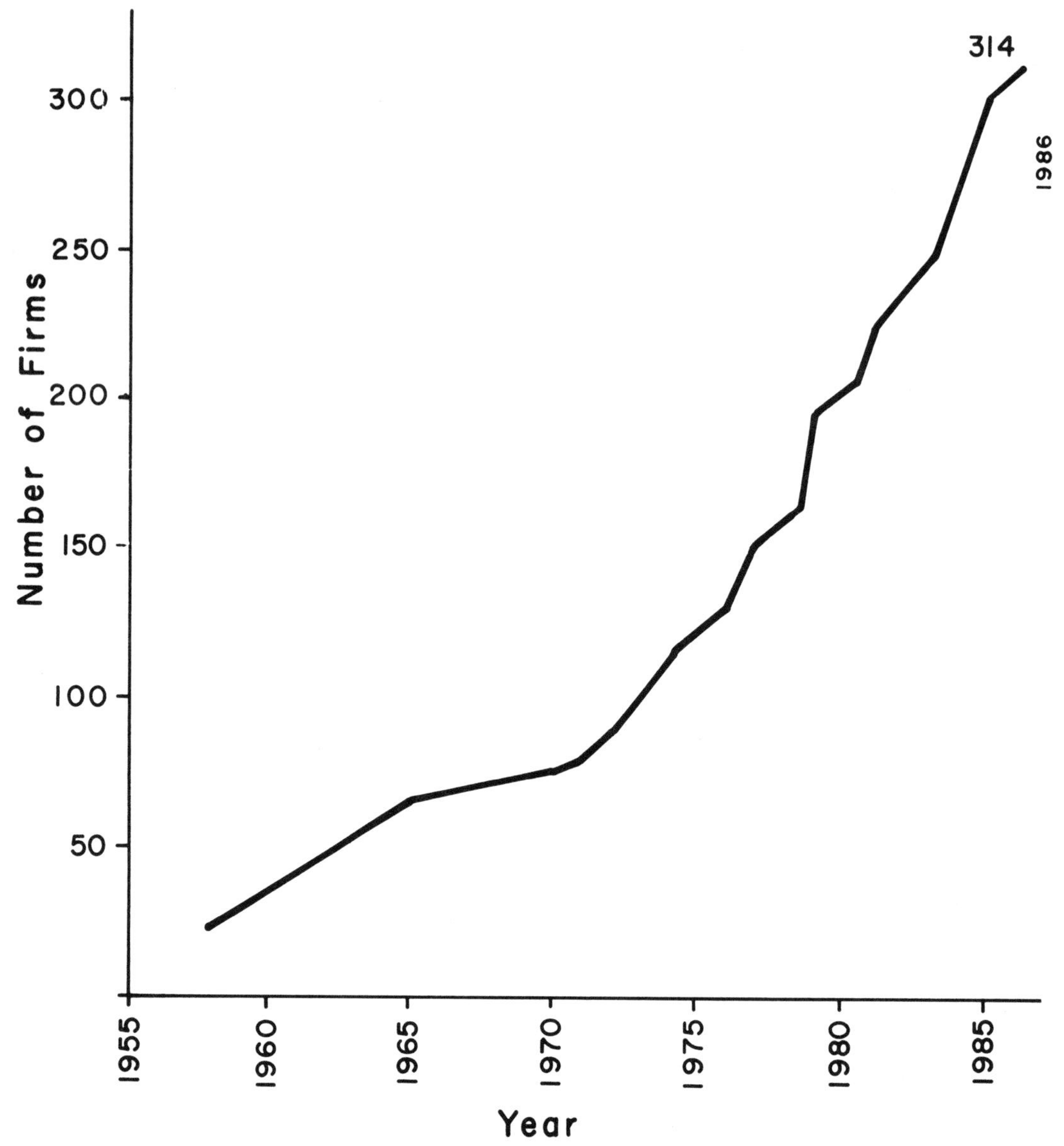


*Even though Colorado's economy and growth slowed down, CECC continued to grow at a strong pace reaching an all time high, towards the end of 1986, of 314 member firms. There were 446 individuals (principals of member firms). CECC firms employ 4,834 people and have gross billings in excess of $250 million per year; an estimated $70 million returning to Colorado from out-of-state work. The annual cost of construction projects arising from the services of CECC firms is over $2 billion. The association's annual operating budget was $186,700.*

# Into the Future

## 1986 - Present

The State in 1986 was alive with the fanfare of political activity. In addition to local and state races, Colorado was getting ready to elect their first new governor in twelve years. Candidates, Ted Strickland and Roy Romer, were not shy about making speeches and attending banquets.

The Council's CEPAC, along with PEC, sponsored one of the many gubernatorial debate dinners and raised $3,263 for both parties. "1986 was one of the most successful in fund raising in recent years," reported Hal Bishop, CEPAC Chairman. He added, "(We) distributed a total of $11,700 to 62 candidates."

President Steve Holt continued CECC's ongoing participation in tort reform. He also orchestrated the negotiations with the State Attorney General on the state's proposed Indemnification Clause for A/E's doing work with the State. The initial language of the clause was impossible for consulting engineers. Holt's efforts resulted in modifications somewhat more palatable.

Building on the process started last year by Larry Muller, Holt, for the first time had CECC committees assigned to five divisions; education, communications and member affairs, business services, governmental affairs and administration, with responsibility for these division activities assigned to specific board members.

*President Steve Holt (left) presents Pat Patrick an "Award of Appreciation" plaque for his many years of service to the Council upon Pat's retirement after 13 years as CECC's lobbyist. Comments Holt, "Pat will remain irreplaceable in the way he has been a friend and of service to us over the years." 1987.*

When Stephen Holt took over as president, he stressed the importance of the liability issue in the consulting business, but at the same time he emphasized "quality work." Steve believed that the ACEC "peer review" program should be actively and rigorously pursued by the membership.

The downturn in the Colorado economy that started at the beginning of the 1980's was lasting longer than most had expected. For the first time in the history of CECC, there was a decrease rather than an increase in membership. "Despite this economic downturn, the vitality of our organization remains quite strong," commented Holt. He continued, "The depth of commitment and resiliency of our organization bodes well for the professional engineering community in coming years."

The legislative committee, with Rich Weingardt as chairman and Andy Andrews and Jack Janney as vice-chairmen headed an energetic group of committee members. Because of their efforts, CECC had significant influence in Colorado's milestone tort reform legislation of 1987.

The committee also actively and aggressively pursued efforts to get more engineers on boards and commissions. A notebook listing all engineers willing to serve was formally presented to Governor Romer by President Holt.

In 1987, Pat Patrick resigned as CECC lobbyist after 13 years of unsurpassed service. He had brought the Council in close contact with legislators. He stated that he was retiring (from his CECC duties) to devote more time to his broadcasting business. One of his many long time acquaintances at CECC, Fu Hua Chen affectionately remarked, "We will miss the quick wit and political 'know-how' of the inimitable Mr. Pat Patrick."

Pat was succeeded by Bill Artist, former state representative from District 48. Bill is president of Artist & Associates, political consultants.

CECC's 1987 Grand Award winner, for engineering excellence, United Concourse "B" at Stapleton International Airport attracted national attention. It was named one of the 12 outstanding engineering projects in the nation by the American Society of Civil Engineers. The structure, engineered by Richard Weingardt Consultants, Inc. was the ASCE/Colorado "Civil Engineering Project of the Year." In 1988, the Associated General Contractors (AGC/Colorado) named it the top Colorado construction project of the year.

Robert Emmons succeeded Holt as the 33rd president of Consulting Engineers Council of Colorado.

During his term, two significant CECC "position papers" were prepared and presented. Appearances were made before the joint highway committee of the legislature to present the Council's position and testify regarding the Arthur Anderson Study of the Colorado Department of Highways; and their use of consultants. John Haley, Al Menhennett and Joe Siccardi were instrumental in preparing the report.

The other major "statement" involved the preparation of CECC's position regarding the professional engineer's licensing law "sunset review." Dick Hepworth was chairman of the committee in charge.

In recognition of his outstanding contributions to consulting engineering, Lee Rice was the fifth man to receive the Orley Phillips award, joining Jim Konkel, Malcolm Meurer, Fu Hua Chen and Orley Phillips.

*Denver Mayor Federico Pena (left) congratulates Richard Weingardt for his firm's achievement in winning the 1987 Grand Award for Engineering Excellence from the Consulting Engineers Council of Colorado. Pena accepted the owner's plaque for the $50 million Stapleton International Airport Concourse B project.*

CECC instigated an awards program to honor outstanding engineers in industry. The task force, headed by Rich Weingardt, named the proposed annual award, the "General Palmer" award, after the founder of Colorado Springs. One of Colorado's greatest "empire builders," William Palmer created the Denver and Rio Grande Railroad, the first narrow gauge line in USA, and was a principal founder of Colorado College and Colorado Fuel and Iron Company. He was known as a Soldier, Engineer, Practical Dreamer, Financier, Philanthropist, and Builder.

The Palmer Award was established as an annual award given to an outstanding engineer in industry who "exemplifies the standards set by General Palmer."

The Long Range Planning Committee, headed by Larry Muller, updated the council's Long Range Plan. The new updated plan was unanimously accepted and approved by the board.

In his "State of the Council" message at his term's end, President Emmons recapped,

"1987-88 has been a year of concern. What had been hoped to be the year of the turnaround did not happen, and we saw businesses in Colorado continue to slide. A new governor has made economic recovery his primary concern, but so far few results can be seen."

*George Wallace, P.E., (shown with his grandson, Jeremy) is the first recipient of the 1988 General Palmer Award, a prestigious award presented by CECC to a Colorado engineer outstanding in industry who has made significant contributions to the state's progress. Wallace, called a visionary by many, is the founder of DTC.*

He closed his address with, "I see major changes happening within our chosen profession in the way we compete for business, the way we accomplish assignments, the way we manage our companies and the way we are perceived by those who use our services."

In his acceptance address at the installation of officers, President Richard Weingardt spotlighted, "Four key issues engineers in Colorado need to deal with are the lack of work in the state, low fees and bidding, frivolous lawsuits and high insurance premiums, and the lack of public understanding of engineers and what we do."

"Consulting Engineers Council," he continued, "is the organization concerned with the business of engineering. Through the Council we can change things, including the future of engineers and Colorado. We can make things happen, control and change our own destiny."

Weingardt established six new ad hoc committees to supplement CECC's standing committees in addressing key issues currently facing consulting engineers in the state. The ad hoc committees, all headed by past presidents of CECC, are: WBE/MBE, statewide coordination, community leadership, new airport liaison, orientation and history — the impact consultants have had on Colorado's progress.

*Gregg Ten Eyck (left), the 1988 CECC "New Principal of the Year," is congratulated by the previous award winner, Dave Stewart. Stewart was also the recipient of the national award by ACEC for 1987.*

Where are we going? What will the consulting engineering business be like in the year 2000? How can CECC best fulfill the future concerns of consulting engineers, and what messages should we be giving our national organization, ACEC, on our needs. These and other society issues were addressed at CECC's first engineer's "think tank" session held at the 1988 Annual State Conference in Steamboat Springs. Al Knott presented an in-depth analysis of society's concepts on standards of care for the professions. Can we ever be a "risk free" society? Expectations that we need to be perfect are causing excessive lawsuits, high insurance costs and deterring creativity and innovation in America. The "life cycle of engineers" is much different in the USA than Japan or West Germany. The rest of the world seems to hold engineers in higher regard as professionals. Can we do anything to change this? The informal brainstorming session was enlightening, thought provoking and highly successful. "Think tanks" by engineers need to be ongoing.

The proposed new Denver airport looms on the horizon as the most important single project that could act as the catalyst improving the state's ailing economy. The new liaison committee, chaired by Mike Barrett, worked closely with Bill Smith, P.E., and others at Denver's New Airport Authority in refining QBS (Quality Based Selection) procedures to be used in hiring A/E's for design work.

Smith, Deputy Director of the Airport Authority, was the May meeting speaker and his presentation attracted a record number of attendees. There will be approximately 80 design contracts let in 1988-89 and they will be chosen by using QBS criteria.

For the first two major design contracts for the new airport, the city hired Bechtel Engineers to prepare the civil engineering standards and the Perez Group to perform

---

### CECC MEMBERS ELECTED TO PUBLIC OFFICE

*Sen. Norton*   *Tallarico*

*Tom Norton, P.E., has been re-elected to the Colorado state legislature. He was first elected to the House of Representatives and is now a Senator in District 48. Bob Tallarico, P.E., is a city councilman with the City of Northglenn.*

similar duties for the architectural standards. In recent months, several additional contracts involving CECC member firms have been undertaken.

Bob Tallarico, chairman of the procurement practices committee, has been very aggressive and vigilant in responding to "Requests for Bids" from any and all governmental agencies attempting to hire engineers by bidding. Under his leadership the committee has been successful in working with Northglenn to develop a QBS Ordinance that is a milestone in the design industry.

Likewise, the transportation committee, under Al Menhennett, is making headway with the Colorado Department of Highways (CDOH) in establishing reasonable profit margins for consultants working for CDOH. The funding of highways and the state's transportation system itself will receive much attention from the governor and state legislators in the coming years.

Construction of the I-25 - Colfax and the I-25 and Sixth Avenue interchanges are quickly reaching completion and will be a tremendous asset to the metropolitan area. The Auraria Parkway system is also now operational which could spur growth in lower downtown and the central Platte Valley.

The legislative committee, under the direction of Larry Muller, with Andy Andrews as Board liaison and Bill Artist as lobbyist, have transportation issues as a priority on the agenda. The committee also continues its active efforts to get engineers on boards and commissions.

The ad hoc committee leadership committee was formed because President Weingardt believes engineers should be leaders; "They should be good engineers, certainly, but at the same time they should play a broader role in society."

"The public," he says, "doesn't think we have the leadership qualities for the big societal issues. Engineers need stronger involvement in educational, professional and civic organizations, and in politics." He emphasized, "Become an expert at something besides your own profession."

Pete Webb was hired as a consultant by the P.R. Committee, which is headed by Gordon Meurer. ACEC was designated the lead association, to chair national activities for Engineers Week, 1989. CECC is similarly the lead group for Colorado activities. Bob Bates is chairman of the local committee. The P.R. committee is assisting Bates and his task force in every way possible to let the public know how "engineers turn ideas into reality," the theme of National Engineers Week.

There are just barely ten years until the year 2000 and many things will change in that time; changes requiring engineering. The updating and repair of our nation's infrastructure — necessary for our basic needs and to maintain our current standard-of-living — will be one of these challenges. Applying the latest technology to new construction, development of space and communications will be others. Engineers will be needed to deal with the "green house" effect and other environmental issues.

Weingardt observes, "As we move into the 21st Century, there will be tough times and good times. Colorado consulting engineers, working hard together with other businesses, can keep the momentum moving forward. By being here we have made a commitment to the vitality of Colorado and the quality of life that makes our state such an attractive place to live and work."

As we approach the future, ACEC President Robert Hogan reminds us, "All of us hope for a better world, but engineers will, in the most literal sense, help build it."

*Fu Hua Chen with president Gerald Ford at his home in Vail, Colorado. 1983.*

# Appendix A

*Williams Fork Dam, Grand County. The dam provides replacement storage for use on the Western Slope of Colorado, to allow transmountain diversions to the Denver Metropolitan area at times when such transmountain diversion would otherwise be out of priority. The original dam is a concrete gravity dam, and the enlargement consists of a concrete arch dam on top of the concrete gravity section. (OW) Board of Water Commissioners, City and County of Denver, (GC) Mountain States Construction Company, (EG) Tipton and Kalmbach, Inc. (1960 photograph).*

# Significant Colorado Projects

Consulting engineers are in many cases responsible for engineering Colorado; for designing the construction projects key to the growth of the state. Members of CECC have had a profound impact on shaping our communities and on the engineering profession itself.

In a recent survey of CECC's membership, the ten most important infrastructure projects constructed in the state since World War II were determined to be:

1. Eisenhower-Johnson Tunnels
2. Interstate 70 (the 450 mile route excluding Glenwood Canyon)
3. Dillon Reservoir and Roberts Tunnel
4. Interstate 25 (the entire 299 mile route)
5. Stapleton Airport
6. Air Force Academy
7. Mile High Stadium
8. Chatfield Dam
9. Glenwood Canyon - I-70
10. NORAD Cheyenne Mountain Complex

Engineers belonging to Consulting Engineers Council of Colorado were involved with many of these projects. The original construction cost for all ten projects was $2.1 billion. To replace them at today's prices would cost $9 billion. Timely investment in infrastructure construction and maintenance is indeed cost effective.

The projects below, though only a partial list, are representative of recent significant Colorado construction engineered by CECC member firms.

*The following abbreviations are used for the projects listed below; architect (AR), civil engineer (CE), engineer-architect (EA), electrical engineer (EE), engineer (EG), general contractor (GC), mechanical engineer (ME), mechanical and electrical engineer (M/E), owner (OW), and structural engineer (SE).*

### ALICE SWEET WATER TANK, ARVADA

The 10 million gallon circular prestressed concrete water tank was winner of 1976 Grand Award for Excellence from CECC; largest circular prestressed concrete tank with restrained base ever built at that time. The project was completed in 1975. (OW) City of Arvada, (EA) Wright-McLaughlin Engineers, (GC) Robert Dougan Construction Co., (SE) Jorgensen, Hendrickson and Close, Inc.

## ARKANSAS VALLEY CORRECTIONAL FACILITY, ORDWAY

One of the first fast-track prison facilities constructed for the Colorado Department of Corrections. Project received national award from Prestressed Concrete Institute. Date completed, 1986. (OW) State of Colorado, (AR) RNL, (GC) G.E. Johnson, (SE) S.A. Miro, Inc.

## AURORA DAM AND RESERVOIR, AURORA

The Aurora Dam and Reservoir is the largest public works project currently under way in Aurora. It will store 30,000 acre-feet of raw water to meet the City of Aurora's peak daily and seasonal water demands.

The dam will be zoned earthfill approximately 8,000 feet long and 137 feet high.

It will be constructed with approximately 5,700,000 cubic yards of onsite borrow materials and about 400,000 cubic yards of imported granular materials. A 106-foot-high multilevel gated reinforced concrete outlet tower will be constructed for selective withdrawal of water for treatment and distribution to the City's water system.

The project also includes over 3 miles of 60-to 72-inch-diameter raw water transmission line, 1 mile of 27-to-30-inch-diameter sewer lines, 3 wetland sedimentation ponds, 2 miles of access roads, and 3-1/2 acres of parking areas. The project, when completed will also feature a park around the reservoir with a marina, several beaches, and a 9-mile bike path. Estimated completion date, January 1990. (OW) City of Aurora, (GC) Industrial Constructors Corp., (EG) CH2M Hill.

## BARRY GOLDWATER VISITOR CENTER, COLORADO SPRINGS

Almost since its inception, more than 25 years ago, the United States Air Force Academy has been the number one tourist attraction in the State of Colorado. However, until the construction of the Barry Goldwater Visitor Center, the handling of these visitors was highly ineffective. The former visitor center, a temporary renovated manufacturing building with Quonset hut additions, was inadequately sized to display to its visitors the mission of the Academy. This building was poorly located, being several miles from the prime cadet academic area. This has all been changed.

The new Visitor Center serves as an attractive and conveniently located entry point from which the visitors can begin their tour of the Academy. This building is sited within walking distance of the renowned Academy Chapel and Academic Overlook area. Its new location now gives an orderly flow of the more than one million tourists who visit the Academy each year. This new building contains an auditorium and a large display area which gives the visitors an overall view of the Academy and explains what the Academy is all about. The center was completed in May of 1986. (OW) United States Air Force Academy, (EG) Leigh Whitehead & Associates, (SE) Howard Dutzi & Associates, (AR) John J. Wallace & Associates, (GC) G.E. Johnson Construction Company.

## BIG SANDY CREEK BRIDGE, HIGHWAY 109, HUGO

The 375' long prefabricated steel bridge is the first multi-span, multi-lane prefabricated steel bridge in Colorado, and perhaps in the U.S. It consists of five 75' spans carrying two lanes of traffic. The superstructure consisting of ten pieces was shipped and erected in three and one-half days. Date completed December 1984. (OW) Lincoln County, (SE) Robert Tallarico, (EG) Matrix Engineers, A.G. Wassennar was the soils engineer.

## BOULDER CREEK GRAVEL EXTRACTION AND ECOLOGICAL RESTORATION PROGRAM, BOULDER COUNTY

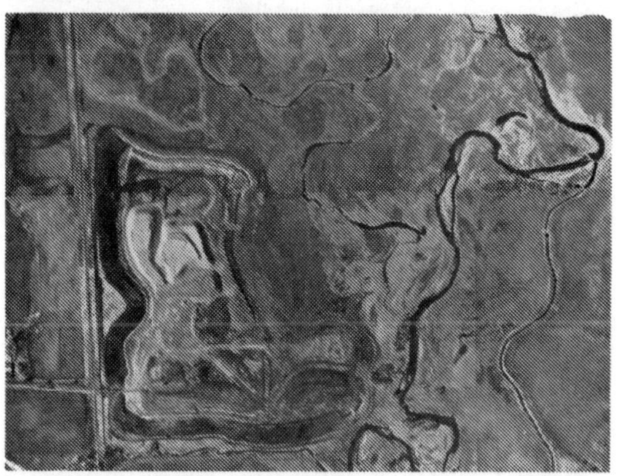

Demonstrated the ability to combine the extraction of a valuable resource in an environmentally sensitive area with a phased restoration program that enhanced the original site and provided a unique ecologically diverse habitat. Date completed, 1974. Won CEC/C and ACEC Engineering Excellence Awards. (OW) Flatiron Sand and Gravel Company, (GC) Flatiron Sand and Gravel, (EG) Leonard Rice Consulting Water Engineers.

## BOULDER WATER SUPPLY TREATMENT AND DISTRIBUTION

Water supply for the Boulder Reservoir water treatment plant is obtained by way of a system of reservoirs, diversions, and canals, all part of the Colorado Big Thompson project. The plant allows the City of Boulder to utilize the reservoir water supply to meet growing water needs. The project included design of water treatment plants, pumping stations, and transmission

lines, as well as comprehensive improvements programs and rate studies. (OW) City of Boulder, (EA) Black & Veatch.

## COLORADO CONVENTION CENTER, DENVER (photo by C.W. Fentress)

Construction on the $125 million state-of-the-art facility is well underway and will be ready for use by April 1990. The 900,000 S.F. concrete and steel structure was designed by C.W. Fentress and Associates, P.C. Associate architects are Loschky, MarQuardt, and Nesholm. Developer is French and Company in conjunc-

tion with Hensel Phelps/Alvarado. (OW) City of Denver, (GC) Phelps/Alvarado, (SE) Martin/Martin with associates H.S. Engineering and Richard Weingardt Consultants, Inc., (EE) Garland Cox and Associates.

## CONCOURSE B, STAPLETON INTERNATIONAL AIRPORT

The project consisted of widening, lengthening and raising the roof on the busiest — 30,000 passengers daily — concourse at the sixth busiest airport in the world, keeping 20 of 22 gates in operation at all times during the entire 18 month construction period. The $48 million project completed four months ahead of schedule and $2 million under the estimated $50 million budget, allowed United Airlines to increase its passenger capacity from 5 million to 10 million travelers per year. It is the recipient of several national as well as local design awards. (OW) City and County of Denver, (AR) Seracuse Lawer Wong Strauch/Roybal and Associates, (M/E) Swanson Rink, (GC) Hensel Phelps/Alvarado, (SE) Richard Weingardt Consultants.

## CURRIGAN EXHIBITION HALL, DENVER

The project features include a space frame roof fabricated on the ground, and a "DIABECK" precast floor system (specifically developed for this project and patented by KKBNA) which saved $0.20 to $0.25 per square foot and over six weeks in construction time. Project won numerous awards from Lincoln Arc Welding Foundation, American Iron & Steel Institute, and the American Institute of Steel Construction. This project was completed

in 1969. (SE) KKBNA, (OW) City and County of Denver, (AR) Muchow Associates/Haller & Larson/James Ream & Associates, (GC) Burkhardt Steel/F.R. Orr Construction Company, (EE) Swanson Rink, (ME) McFall/Konkel.


## DENVER/BOULDER TURNPIKE, HIGHWAY 36-DENVER TO BOULDER

The first and only toll highway in the state. The project was so successful that bonds were retired and tolls removed in 1967. The highway is currently one of the important elements of the metro Denver transportation system. Date completed, 1952. (OW) Colorado Department of Highways, (EG) Howard Needles Tammen & Bergendoff.


## DILLON DAM, SUMMIT COUNTY

A 231-foot high and 5,888-foot long embankment dam. Dillon Reservoir is one of the major reservoirs located on the Western Slope for storing water originating on the Western Slope for subsequent transmountain diversion to the Denver metropolitan area. Planning studies for obtaining water from the Western Slope

*(1963 photograph)*

were started in 1913, construction of Dillon Dam was started in 1959, and construction was completed in 1963. (OW) City and County of Denver, (GC) Potoshnick Construction, Inc., (EG) Tipton and Kalmbach, Inc.


## FOOTHILLS WATER TREATMENT PLANT, JEFFERSON COUNTY

The Foothills Water Treatment Plant was designed to be built in four increments of 125 mgd. The first increment (Phase I) was completed in 1983. The second increment of construction (Phase II) was completed in early 1986, bringing the plant to a nominal capacity of 250 mgd. The plant has been designed with the flexibility to operate either in the complete treatment mode or in an in-line filtration mode. This accommodates variations in influent water quality.

Although initial construction, including the cost of the hydroelectric generation facilities, was a substantial sum, it is equivalent to only $420,000 per mgd of nominal capacity. Phase I included facilities for the ultimate plant development of 500 mgd, including yard piping, the chemical field, and the chlorine building.

Phase II of the Foothills Treatment Plant construction increased the nominal capacity of the plant by 125 mgd at a cost of only $12.5 million dollars or $100,000 per mgd. Costs to date: $65 million per mgd.

The plant has proven its capability to consistently produce high-quality water at rates exceeding the nominal rated capacity. It has also proven to be very economical to operate, with total operation and maintenance costs for 1986 projected to be less than three cents per thousand gallons of water produced. Phase I was completed in 1983 and Phase II in 1986. (OW) Denver Board of Water Commissioners, (EA) CH2M Hill, (GC) Phase I, Martin K. Eby; Phase II, Madsen Western Builders. DMJM provided structural design of treated water storage reservoirs.

## GLENWOOD CANYON TUNNELS, GLENWOOD

The Glenwood Canyon Tunnels posed a special challenge for both design and construction because the project lies in an environmentally sensitive area which has enjoyed intense community use since the turn of the century. Controversy over the project was resolved by the decision to place the highway in twin tunnels through one wall of the canyon, parallel to the river, through the most sensitive area.

Solutions to difficult geotechnical and construction problems had to be coordinated since the portals are located in unstable areas on slopes where rock slides could potentially occur. An exploratory tunnel led to the design of grouted rock dowel support systems — initially recommended by Parsons Brinckerhoff — which reduced heavy construction costs

by 35 percent. Both of these major highway tunnels utilize state-of-the-art tunnel management and traffic control system. Estimated completion January, 1989. (OW) Colorado Department of Highways, (GC) J.F. Shea Company, (EG) Parsons Brinckerhoff, Quade & Douglas, Inc.

## GREEN MOUNTAIN EXCHANGE PROJECTS, WESTERN COLORADO

This two-year study confirmed the feasibility of developing an additional water supply for the Denver Metropolitan area averaging 120,000 acre-feet per year without constructing major diversion systems in environmentally sensitive areas of the Western Slope. The new water supplies would be collected in Green Mountain Reservoir, pumped back up the Blue River Valley to Dillon Reservoir, and conveyed to the East Slope through the existing Roberts Tunnel. This concept, which was suggested by Governor Lamm's Metropolitan Water Roundtable in 1982, involves replacing the present function of Green Mountain Reservoir by a new reservoir or reservoirs in the Colorado River basin in western Colorado. Date completed, April, 1987. (OW) Colorado Water Resources & Power Development Authority, Soils Engineer was Chen & Associates, (EG) Boyle Engineering Corp.

## GREEN MOUNTAIN RESERVOIR, LAKEWOOD

The 30-mgd pumping station and the five-million-gallon underground water storage reservoir were built to boost treated water from Denver's system into Lakewood and the rest of the Green Mountain Water and Sanitation District's service area. The Green Mountain facilities serve a population of approximately 125,000. The facility features a cover of contoured earth over both structures and exposed structural concrete. The station has a maximum pumping capacity of 30 million gallons per day. It is equipped with four horizontal centrifugal pumping units. The future maximum pumping capacity can be increased to 40 mgd. (OW) Denver Water Board, (EA) Black & Veatch.

## GREYHOUND/TRAILWAY BUS TERMINAL REHABILITATION, DENVER

Rehabilitation of upper level (rock) parking decks and ramps. This is the state's largest private bus facility and the repairs should significantly extend the useful life of this facility. The latest concrete repair technology and products were utilized to include polymer modified concrete toppings; epoxy-resin non-shrink grouts; sealants; and elastomeric waterproofing membranes. This project was completed in August, 1988. (OW) Greyhound Lines, (EG) Arnesen & Associates, Inc., (GC) Artesian Inc.

## GROSS DAM AND RESERVOIR OUTLET WORKS, DENVER

South Boulder Creek Canyon, 35 miles northwest of Denver, is the site of the new outlet works for Gross Dam and Reservoir which provides a portion of the water demands for the City. Discharge from the outlet works flows to Ralston Reservoir and then to the Moffat Treatment Plant for treatment and distribution. The dam, reservoir, and outlet works were completed in 1954. Gross Reservoir is contained by a concrete gravity arch dam standing 330 feet in height above South Boulder Creek. The reservoir has a capacity of 43,064 acre-feet of storage. (OW) Denver Board of Water Commissioners, (EA) Black & Veatch.

## GROUND ACCESS PLAN FOR NEW DENVER AIRPORT, DENVER

This facility is proposed to be the world's largest airport to handle domestic and international flights. The ground access requirements had to address convenient and rapid connections to metropolitan and state freeway facilities, as well as key surrounding arterial systems. Date completed, December, 1988. (OW) City of Denver, (EG) Felsburg Holt & Ullevig.

## I-25/C-470 INTERCHANGE, DENVER

First phase of the directional interchange provided connection of C-470 with I-25. The structures are designed for second phase construction that will complete the interchange as a part of the E-470 project.

Date completed, 1986. (OW) Colorado Department of Highways, (GC) Flatiron Structures, (EA) Howard Needles Tammen & Bergendoff.

## I-25/COLFAX INTERCHANGE AND COLFAX VIADUCT, DENVER

Reconstruction and improvement of one of Denver's major interchanges. As an integral part of the interchange, replacement of the Colfax Viaduct was accomplished at a cost of $31 per square foot. Date completed 1984. (OW) Colorado Department of Highways, (GC) Centric Corporation for Colfax Viaduct portion, (EG) URS Company, (EA) Howard Needles Tammen & Bergendoff.

## I-70 GLENWOOD CANYON, GLENWOOD CANYON

The Glenwood Canyon project is one of the last parcels of I-70 to be completed, but it is most significant to Colorado. Complexity of the project resulted in the need for innovative construction techniques never used in U.S. highway construction. Environmental, physical, operational and funding constraints have resulted in numerous award-winning designs and traffic control methods that have gained international attention. The project has addressed all of the concerns of numerous special interest groups. The French Creek Viaduct consists of two precast concrete segmental bridges, 1300 feet and 3100 feet in length.

Numerous CECC member firms have worked on the planning, design and construction management of the project which began with design concepts in 1974. Prominent among these engineering firms are DMJM, Deleuw-Cather, HNTB and Centennial. Several contractors have also worked on the many phases of construction including Flatiron Structures, Kiewit Western, Centric/Jones, Tectonic, J.F. Shea, Shank-Artukovich and Ames Construction. (OW) Colorado Department of Highways.

## LITTLE DRY CREEK FLOOD CONTROL IMPROVEMENTS, ENGLEWOOD

The City of Englewood contracted with McLaughlin Water Engineers, Ltd. for design of improvements to relieve an estimated $14 million in potential damage from a 100-year flood along Little Dry Creek, then a debris-strewn eyesore. The project was designed to increase the city's safety and economic viability by taking downtown areas out of the floodplain and providing a linear park link to the upstream community. Three project elements work together to accomplish these goals; the downtown plaza area combines flood conveyance with an on-stream lake, urban plazas and water features that define the city's center; the channel improvements

provide flood conveyance and over 8 acres of open park space; and and the off-channel storage facility provides both 85 acre feet of flood storage and a 14-acre athletic complex featuring two soccer fields and a baseball field, complete with pressbox and bleacher facilities. Construction cost: $8.6 million. Year completed: 1988. (OW) City of Englewood and Englewood Urban Renewal Authority, (GC) Palisade Construction, Randall & Blake, and Lillard & Clark Construction, (EG) McLaughlin Water Engineers.

## LITTLETON RAILROAD DEPRESSION, LITTLETON

The project improved traffic operations and safety by eliminating at-grade railroad crossings at four arterial streets and replacing them with three crossings of the depression. Nine thousand feet of Denver & Rio Grande Western Railroad (D&RGW) and Atchison, Topeka and Santa Fe Railway (AT&SF) track was depressed while maintaining rail traffic and vehicle traffic on arterial streets throughout construction. Value engineering analysis was applied to key aspects of the project and was successful in reducing the cost from the originally estimated $20 million to $17 million.

Five thousand linear feet of mechanically stabilized embankment (MSE) retaining walls up to 35 feet tall were constructed. MSE walls are not typically used continuously along railroad tracks. The MSE system was selected for the project to meet the geotechnical requirement to allow for potential movement due to locked in stresses from over-consolidated soils and to effectively collect and convey drainage water.

The concrete flume crossing the Depression is capable of carrying 2,700 cubic feet per second. It has a 25 foot stair-step drop structure at its outlet to the Little's Creek channel. Its design was adapted from a Bureau of Reclamation dam spillway, and was modeled at the Bureau's Engineering and Research Center in Denver. (OW) Colorado Department of Highways/City of Littleton, (EG) Centennial Engineering. The project was completed in the Spring of 1988.

## LOWER SACRAMENTO CREEK RESERVOIR, PARK COUNTY

This small water rights augmentation reservoir was built for the joint use by more than five mountain communities. It is an off-channel reservoir built on a glacial deposit. A number of attempts were made to seal the reservoir from excessive leakage with poor to fair results. Excessive leakage resulted in all the communities' plans for augmentation being out of compliance with their water court decrees. An engineering plan was developed to quantify and claim the reservoir leakage for credit as replacement water under their plans for augmentation. This plan was submitted to the Water Division 1 Water Court and was approved after thorough review by other users who objected to the plan. Date completed, 1985. (OW) Lower Sacramento Creek Reservoir Company, (EG) Blatchley Associates, Inc.

## MASONIC BUILDING WALL BRACING & RECONSTRUCTION, DENVER

Preserved and restored to useful service a historic building in downtown Denver after fire completely gutted the interior leaving only the exterior 7-story high brick and stone masonry walls standing. The walls were braced from the outside allowing removal of debris and reconstruction to take place on the interior. Date completed 1985. (SE) Borman/Smith/Bush & Partners, (OW) Cambridge Development Group, (AR) Gensler & Associates, (GE) G.E. Johnson.

METROPOLITAN DENVER SEWAGE DISPOSAL (ORIGINAL) PLANT, COMMERCE CITY

At the time it was constructed the facility was the largest single construction contract in the State of Colorado, awarded at $17.3 million.  Also it was the largest wastewater treatment plant (117 mgd), the major wastewater treatment plant in Colorado and intermountain west.  HDR, as a member of Denver Metro Engineers, a three-firm joint venture which also included Phillips, Carter, Osborn, and the Ken R. White Co., all of Denver, designed this facility.  Project completion was in 1967. (OW) Metropolitan Denver Sewage Disposal District No. 1, (EA) HDR Engineering, (GC) Mead and Mount Construction.

MILE HIGH STADIUM, MOVABLE EAST STANDS, DENVER

The Movable East Stands is 450 feet long, 200 feet deep, 135 feet or 13 stories high, seats 21,000 people and weighs nine million pounds.  It is the largest structure ever to be moved by the "fluid film" water-bearing system.  The structure can be moved forward or back 145 feet to accommodate both football and baseball.  It was the recipient of three national engineering awards for outstanding achievement in

1978, including the ACEC Honor Award. The project was completed in June of 1977.  (OW) City and County of Denver, (EA) DMJM-Phillips, Reister, Haley, Inc., (GC) Petry-CM, (M/E) Swanson-Rink and Associates.

N-S RUNWAY AND TAXIWAY STRUCTURES OVER I-70, U.P.R.R & SAND CREEK, STAPLETON INTERNATIONAL AIRPORT

These structures were the first of their kind carrying aircraft over other modes of transportation.  They now have the ability of carrying a one million pound aircraft. These bridges have been in service for 25 years carrying millions of tons of load.

The upgrading of the original structures won the CEC and <u>Engineering News Record</u> awards of 1978.  The original

structures were completed in 1962 and the upgrading in 1977. (OW) City and County of Denver, (GC) A.S. Horner Construction, (SE) Meheen Engineering Corporation, and Chen and Associates and Woodward-Clyde Consultants were soils engineers.

## ONE DTC, DENVER TECHNOLOGICAL CENTER

This state-of-the-art highrise recently received another in a long list of design awards; the 1988 Greenwood Village Architectural Award. Comment the judges, "the buildings form created a desire for the pedestrian to enter." They added, "One DTC, 14 stories of polished granite, polished stainless steel and glass rises above the ground on columnar forms to allow the natural land forms to flow beneath the building. The alternate mix of materials forms a ribboned curvilinear

sweep that gleams against the sky." The structure for the "elegant contemporary" style architecture is a refinement of the "Denver Frame," a structural system so named because it originated in Denver. Gravity loads around the core carry the

gravity loads but instead of concrete shear walls large structural steel x-bracing in core walls resist lateral loads. Completed in 1986 at a cost of $20 million. (OW) Corum Real Estate, (AR) C.W. Fentress and Associates, (GC) G.E. Johnson, (SE) Richard Weingardt Consultants.

## PIKES PEAK CENTER AUDITORIUM, COLORADO SPRINGS

The building of this auditorium was a turning point for the performing arts in Colorado Springs. It had a major impact on the Colorado Springs Symphony Orchestra — its attendance almost doubled when the building opened. It tripled the total number of theatrical performances — dance, opera, musicals, drama, etc. — in the Pikes Peak area. Its acoustics has been highly acclaimed by such noted performers as Leontyne Price and Isaac Stern. This hall was visited by representatives from Dallas and Philadelphia, and its success inspired them to use this hall as a model for the concert halls which are currently being designed for their respective cities. The project was completed in October of 1982. (OW) El Paso County, (AR) John J. Wallace & Associates/Clifford Nakata & Associates, (GC) G.E. Johnson Construction Company, (SE) Howard C. Dutzi & Associates.

## RALSTON CREEK BRIDGES AND CHANNEL IMPROVEMENTS, ARVADA

The project streamlined channel flow and dramatically improved traffic conditions at Wadsworth and three other major streets over Ralston Creek. Included along with channel improvements were bike and walking trails, retaining walls and four concrete box girder bridges. Completed 1981 at a cost of $5.5 million. (EG) Hydro-Triad/Richard Weingardt Consultants, (OW) City of Arvada.

## RAY D. NIXON POWER PLANT WATER SYSTEMS, COLORADO SPRINGS

The zero-discharge water management system for the 200-megawatt coal-fired Ray D. Nixon Power Plant is an integrated water supply and wastewater treatment system designed to meet the power plant's needs for water (4 mgd) and at the same time satisfy strict wastewater discharge standards. Major features of the closed-loop system are: a diversion dam, a rehabilitated well field, a three-mile steel water supply pipeline, an equalization basin, a 1-mgd effluent treatment plant, and solar evaporation ponds for waste brines.

Ninety-five percent of the water is recovered for reuse in the power plant, eliminating the need for a discharge permit. The remaining brine slurry, with a salt concentration three times that of seawater, is separated into liquid and solid components. The residual salts are collected and disposed of in a sealed landfill, and the liquid is pumped to lined solar evaporation ponds. The total water system was completed at a cost of $12 million, and the total power plant cost was $110 million. (OW) City of Colorado Springs, (EG) CH2M Hill.

## HAROLD D. ROBERTS TUNNEL, SUMMIT AND PARK COUNTIES

*(1960 photograph)*

A 23.3-mile long water transmission tunnel with finished diameter of 10'-3, this tunnel is one of the major transmountain water diversion tunnels, bringing water from the Western Slope of Colorado to the metropolitan Denver area. A filing for the water right was made in 1927, limited construction was carried out from 1946 to 1956, and the major part of the work was performed by Blue River Constructors between 1956 and 1962 — 35 years from filing to completion. The final inspection was in May, 1962. (OW) City and County of Denver, (GC) Blue River Constructors, (EG) Tipton and Kalmbach, Inc.

## RTD BUS MAINTENANCE FACILITY - EAST METRO DIVISION, AURORA

This is the largest air-based solar space-heating system in the world. It set a new standard for design and reliability of very large solar systems of this type. Date completed, 1980. (OW) RTD, (AR) RNL, (GC) Hensel Phelps, (EG) Jan F. Kreider & Associates.

## RTD BUS MAINTENANCE FACILITY - PLATTE DIVISION, DENVER

This is the largest liquid-based solar space-heating system in the world. It set a new standard for design and reliability of very large solar systems of this type. Date completed, 1978. (OW) RTD, (AR) Charles Sink, (GC) Hensel-Phelps, (EG) Jan F. Kreider & Associates.

## 16TH STREET TRANSITWAY MALL, DENVER

The project has been nationally recognized for revitalizing downtown Denver and creating a memorable urban space. This project was completed in 1981. (OW) City of Denver and RTD, (AR) I.M. Pei & Partners, and KKBNA were civil and structural engineers.

## SNOWMASS SEWAGE TREATMENT PLANT, SNOWMASS VILLAGE

Master planning of the new community of Snowmass Village resulted in the sewage treatment plant being located in a low visible meadow, surrounded by the golf course and high-value development. The plant had to be aesthetically pleasing, have no odors, handle highly variable winter peak flows — and produce a consistent high quality effluent suitable for golf course irrigation or discharge to a trout stream. The answer was a completely enclosed sod-roofed facility. The success of this innovative design — the first known totally enclosed sewage treatment plant in the United States — has influenced many newer treatment plants in cold weather mountainous areas. Initial construction cost: $329,000. Year constructed: 1967, expanded in 1984. (EG) McLaughlin Water Engineers, (OW) Snowmass Water and Sanitation District, (AR) Russ Peilstick, (GC) Utilities Engineering and Construction Company, (ME) Rice-Marek & Associates, (EE) C.H. Hoper & Associates, (SE) Jorgensen & Hendrickson.

## SOUTH PLATTE RIVER GREENWAY, DENVER

The South Platte River Greenway in Denver has transformed an open sewer into a community amenity, resulting in increased property values and a new civic pride. Flood damages are reduced, parks are available for citizens and the pathway to improved development now exists. (Confluence Park, an unusual City park, is shown in the photo.) This project was completed in 1985. (OW) The Greenway Foundation and The City of Denver, (SE) Jorgensen, Hendrickson & Close, (EG) Wright Water Engineers.

## STAGECOACH DAM AND RESERVOIR, ROUTT COUNTY

Provides multi-use of Colorado water for: recreation, agricultural, municipal, industrial, wildlife and fish enhancement, and hydroelectric power. This project was completed in December, 1988. (EG) Woodward-Clyde Consultants, Tudor Engineering Company, CDC Inc., (OW) Upper Yampa Water Conservancy District, (GC) ASI-RCC, Inc.

## STAPLETON NOISE INSULATION PROGRAM (SNIP), DENVER

This project involved the design of noise reduction for more than 3,200 homes, 20 churches and 8 schools impacted by aircraft operations at Stapleton International Airport. Design completed December, 1987, construction completion projected for December 1990. (OW) Colorado home owners, churches, Denver Public Schools, University of Denver (CWC Campus), Aurora Public Schools, (AR) W.C. Muchow & Partners, (EG) David L. Adams Associates.

## STELLAR PLAZA, DENVER

Located in downtown Denver, Stellar Plaza is a mixed-use complex that offers a unique living and working environment.

Integrated by a distinctively landscaped urban plaza, the complex includes a 37-story hotel apartment tower, a 31-story office tower and three levels of underground parking.

Responding to the client's needs and DURA's (Denver Urban Renewal Authority) requirements, and the connecting plaza were elevated above street level to increase pedestrian circulation. (SE) KKBNA, (EE) Garland Cox, (ME) BHCD, Architect/Landscape Architect: Seracuse Lawler Wong Strauch.

## STERLING WASTEWATER TREATMENT PLANT

The innovative feature of this project is a groundwater recharge system that will provide the City with increasing water right credit into the next century. The project was one of seven Grand Award winners in the ACEC 1983 competition, selected as the top environmental project nationally. This project was completed in 1982. (OW) City of Sterling, (EA) Arix Corporation, (GC) Western States Construction Company. Resource Consultants, Inc. provided groundwater hydrology and Chen & Associates provided geotechnical services.

## STRONTIA SPRINGS DAM, JEFFERSON & DOUGLAS COUNTY

A state-of-the-art computer analyzed high concrete arch dam reaching a height of 300 feet . Siting of the project allows Denver Water Department opportunity to supply water by gravity to most of the systems and eliminate pumping. Small hydroelectric power plant was also designed and added to dam to recover wasted water power from required water releases to maintain fish flows and meet water rights obligation. Project completed in 1983. (OW) Denver Water Department, (GC) Morrison-Knudsen, (EG) Harza Engineering Co., JF Sato was associate engineer, DMJM designed tunnels & CH2M Hill designed Foothills Treatment Plant.

## TACOMA HYDROELECTRIC PROJECT, DURANGO

Complete renovation of an existing hydroelectric project. The project included a 22,000 acre-foot reservoir formed by construction of a rock fill dam with a hydraulic asphalt lining on the upstream face; about 9,600 feet of 66-inch diameter low-pressure penstock, supported by ring girders on 80-foot spans; and a 54-inch, high-pressure penstock (design pressure 470 psi), installed on grades of up to 80 percent. Project completed in 1982. (EG) W.W. Wheeler and Associates, (OW) Colorado-Ute Electric Associates, (GC) Nielson, Inc., and Chen-Northern, Inc. were soils engineers.

## UNION COLONY CIVIC CENTER, GREELEY

It was a community effort from the beginning with much of the funding coming from donations from private citizens and business firms in the Greeley area. It has state of the art acoustical features. (OW) City of Greeley, (EA) Arix, (GC) Hensel Phelps Construction Company, (EG) David L. Adams Associates.

## UNIVERSITY OF COLORADO ENGINEERING SCIENCE CENTER, BOULDER

This was one of the first major expansion projects of the University. It features extensive use of exposed architectural concrete. Project completed in 1964. (OW) University of Colorado, (AR) Associated Architects of Colorado (Fisher & Davis - Muchow - Ketchum & Konkel), (GC) Dan R. Ponda Company, (EE) Swanson-Rink, (SE) Ketchum & Konkel, (ME) McFall-Konkel & Kimball Consulting Engineers.

## VAIL PEDESTRIAN BRIDGE, ACROSS INTERSTATE 70 AT VAIL

This bridge provides the only pedestrian access between the main Town of Vail and the residential areas north of I-70, and links the town directly to the new elementary school built at the same time. It has a unique architectural expression with distinct curvature. It was built with long arched box girder sections erected on shoring in the medians without closing the highway, and post-tensioned together to produce the final long-span structure. Date completed, 1977. (OW) Town of Vail, (AR) Pierce, Briner, Fitzhugh, Scott, (GC) Centric, (SE) Johnson-Voiland Archuleta, and Chen and Associates were the soils engineers.

## VAIL PUBLIC LIBRARY, VAIL

Largest fully earth sheltered institutional building in the state and the only library in the west. Earth sheltering was used to reduce energy costs and mitigate visual impact on a picturesque site midway between Vail Village and Lionshead Village on Bear Creek. The library was completed in June of 1983. (OW) Town of Vail, (AR) Snowden, Hopkins, (GC) Kyburz Construction Company, (SE) Jerry York.

## WELL PUMP STATION A-5A, ARAPAHOE COUNTY

At the time, the project involved installation of the largest variable speed submersible water well pump in the state. The pump has a 600 hp motor capable of 1500 gpm from 1000 feet. A variable speed drive is used to control the pump output,

matching water demand in the District and saving power costs. This project was completed in February, 1987. (OW) Willows Water District, (EG) Meurer & Associates.

## ZUNI FACILITY FOR RADIO AND DATA PROCESSING, DENVER

At the time it was constructed it was the largest data processing center west of St. Louis. It is the hub telephone relay station for Denver. This project was completed in 1982. (SE) Anderson & Hastings, (OW) U.S. West and AT & T, (AR) C.B. Guy, (G.C.) G.H. Phipps, (ME) Paul Koch, (EE) Swanson-Rink.

# Appendix B

# CECC Charter Members

**STRUCTURAL ENGINEERS:**

Frederick G. Anderson

Rhuel A. Andersen

Hale V. Davis, Jr.

Clifford Johnson

Ib Falk Jorgensen

E. Vernon Konkel

Milo S. Ketchum

Robert S. Nedell

Orley O. Phillips

Joseph W. Sallada

Norman H. Todd

Kenneth R. White

**ELECTRICAL ENGINEERS:**

Robert V. Behrent

John E. Tracy

Albert E. McKittrick

Frederick J. Rink

Alvin D. Swanson

Arthur F. Weers

**MECHANICAL ENGINEERS:**

Harold W. Marshall

James J. Johnson

Kenneth J. Murray

Arthur M. Riley

Robert L. Whittlesey

Rupert J. Fooks

Maurice S. Wilson

Paul H. Koch

Earl L. Heckman

Francis E. Stark

James H. Konkel

William W. Trego

# Profiles of CECC Presidents

## 1955-56 MURRAY

KENNETH J. MURRAY was born in eastern Illinois in 1905. At an early age he went to work in the sales and engineering departments of American Air Filter, Inc. During World War II he worked in the engineering department of the Grand Island, Nebraska Ordinance Plant. He opened a consulting mechanical engineering office in Denver, Colorado in 1951. Projects on which he provided the mechanical engineering include the Ellsworth Air Force Hospital at Rapid City, South Dakota; the Aeronautical Training Center, Cheyenne, Wyoming; the Science Building at Lamar, Colorado Junior College and the St. Joseph Hospital in Del Norte.

He was very active and instrumental in organizing the Consulting Engineers Association of Colorado (CECC) and served as the first president during the 1955-56 term. Prior to his death he received an award for his services rendered to the Association. He was a charter member, from the Colorado association, of the national Consulting Engineers Council. He retired in 1960 and died in 1970 leaving his wife, May, and two children, Barbara and Robert.

## 1956-57 KETCHUM

MILO S. KETCHUM. With an unusual display of good judgment, Milo was born in Denver, Colorado on March 8, 1910. In six months, however he moved back to Boulder where his father was Dean of Engineering at the University of Colorado. His early years were spent playing in the irrigation ditches on the CU campus and listening to the reports of the War, highlighted by a flu epidemic and the Armistice Day celebration. In 1919 the family moved to Philadelphia, and in 1922 to Urbana, Illinois, where his father again became a Dean of Engineering.

Having no other reason than that it was good training, Mr. Ketchum enrolled in the course of Civil Engineering at the University of Illinois. He soon developed a passionate interest in the design of airships, even inspecting the construction of the USS Akron, under construction in Ohio. This field of engineering, however, soon disappeared with the loss of the Hindenburg and the two US naval airships.

We find our hero, in 1934 again in Colorado, with the Bureau of Reclamation, then in Chicago, and next, in 1937, teaching at the Case Institute of Technology, in Cleveland. It was there that he met and married the beautiful, talented, and intelligent Gretchen Allenbach, who became the real brains of the team. In 1945 he moved to Denver, and opened an office for the practice of structural engineering.

His creative career is divided into three phases. 1) The design of steel rigid frames, during which many unusual buildings were constructed, mostly for schools. 2) His concrete shell structures period, thanks to the cooperation of his architectural clients. This led to speaking engagements all over the country and a national reputation in this field. 3) Four beautiful and intelligent children, David, Marcia, Matthew, and Mark. The last was also to become a structural engineer, now with a PhD, working for a consultant in San Francisco.

In 1962, feeling the desire to explore new territory, he moved to Connecticut to open a branch office of the firm. One friend wrote a letter addressed as follows: "Old Milo Ketchum, Old Boston Post Road, Old Saybook, Connecticut 06475." In 1967, Mr. Ketchum turned the office over to the resident partner, quit work, and went to teach at the University of Connecticut. He retired from teaching in 1978, and resides now at 165 Estes Street, Lakewood.

During the course of his career, he wrote many papers, attended national meetings of societies, and managed to accumulate a number of friends with the result that he received the following awards: Honorary Doctors Degree from the University of Colorado, 1976; The Turner Medal of the ACI; Honorary Member of the ACI; Honorary Member of the ASCE and Member of the National Academy of Engineering.

*(The above is all certified to be true, because it was written by Milo S. Ketchum and corrected by G.A. Ketchum, October 1988.)*

### 1957-58 ANDERSEN

RHUEL A. ANDERSEN was born in Fremont, Nebraska and received a B.S. in Civil Engineering from the University of Nebraska in 1931. Prior to establishing his own consulting firm in Denver in 1947, Andersen worked with the Tennessee Valley Authority, U.S. Corps of Engineers where he was head of the design section in responsible charge of the Cherry Creek Flood Control Dam and with the U.S. Bureau of Reclamation where he was Senior Engineer in charge of structural design for the substructure of the Grand Coulee Pumping Plant.

His firm and the successor firm of Andersen, Koerwitz & Hawes performed structural engineering services for many notable buildings in Denver and the state of Colorado including Manual High School, State Services Office Building, Denver Art Museum and First National Bank and parking structure.

Andersen was a member of Denver Structural Engineers Association, one of the founding members and third president of the Consulting Engineers Council of Colorado, served on the National Board of Directors of Consulting Engineers Council where he was chairman of several committees, was a Fellow and Honorary Life Member of the American Consulting Engineers Council and Life Member of American Society of Civil Engineers. Rhuel is survived by his wife, Ellen, a daughter, Alice Schroeder, and two grandchildren.

## 1958-59 KOCH

PAUL H. KOCH was born in Bee, Nebraska in 1918. During World War II he served with the Aviation Engineers in the Pacific Theater. Following the war he completed his mechanical engineering degree at the University of Nebraska in 1947. After working as a design engineer five years for Marshall & Johnson and two years for K. J. Murray, he opened his consulting mechanical engineering office in 1954. Notable construction projects include the mechanical design for the Denver U.S. Court House & Federal Office Building, the University of Utah Medical Center, The Denver Mountain Bell Equipment and Computer Complex and various schools, hospitals and telephone buildings throughout the Mountain States.

He served several years as an arbitrator for the American Arbitration Association. He was the fourth president of CECC and is a Life Member of CECC and ASHRAE. He obtained the grade of Fellow in the American Society of Heating, Refrigerating and Air Conditioning Engineers in 1978. After specializing for ten years in energy conservation work, he retired in 1984. He has a wife, Alby and two sons.

## 1959-60 SWANSON

ALVIN D. SWANSON was born in Osceola, Nebraska in 1922 and came to Denver fourteen years later. A South High graduate, he served in the Navy during World War II, where he earned his electrical engineering degree from the University of New Mexico in 1946. After working for General Electric in Ft. Wayne, Indiana for a few years, the Swanson family returned to Denver where Al worked for an electrical contractor during 1948-49, followed by establishing his electrical consulting engineering firm in October, 1949. That firm became Swanson-Rink and Associates in 1952, which Swanson led, as its president, until his departure in 1984 to form Engineering Management Consultants.

He received a Distinguished Engineering Award from the University of Colorado College of Engineering in 1975. Following his many years of distinguished leadership in professional and civic affairs, Al suffered a lengthy illness and died in July, 1988, leaving his wife Bonna, his daughter Lynettte, and his son Brion.

## 1960-61 TRACY

JOHN E. TRACY graduated with honors from Rose Polythnic Institute in 1941 with a Bachelor of Science in Electrical Engineering. During World War II he served in the U.S. Army Corps of Engineers from 1942 to 1945. He was then employed as an electrical engineer by Delco Products Division of General Motors. Two years later he accepted a position as assistant plant engineer for the Standard Register Company of Dayton, Ohio. Here he served in a supervisory capacity over the maintenance department and had charge of electrical and mechanical design of a large plant expansion program. Tracy moved to Denver, in 1951, where he was employed by The Austin Company as an electrical engineer. He was in charge of a design group for the electrical installations for a number of administrative and process type buildings at the Rocky Flats Plant.

In February, 1953 he became co-founder with Bob Behrent of the Tracy-Behrent Engineering Company, serving as the firm's president for many years. He was a registered professional engineer in Colorado, Wyoming and New Mexico and was a member of the Professional Engineers in Colorado as well as CECC.

## 1961-62 WAGGONER

EUGENE B. WAGGONER was born in Kansas City, Missouri in 1913 and moved with his parents when he was seven years old to the Los Angeles area. He attended UCLA where he earned a master's degree and a bachelor's degree in Geology. His first work was in the oil fields with Tidewater-Associated and later with Superior Oil in Colorado. When the Reclamation Bureau opened up its engineering geology department, he joined the Bureau. In the ten years of working there, he travelled to a great number of countries.

In 1956 he formed his own consulting practice and after a few years he joined Bill Clevenger as co-partner in the Denver Woodward-Clyde consulting firm. During this time he became national president of Consulting Engineers Council, the first member of CEC Colorado to become president of the national association. In 1967 he became president of Woodward-Clyde Consultants with headquarters in San Francisco. He provided consulting services for the St. Lawrence Seaway, several dam projects in Venezuela and the Bhumaphol Dam in Thailand.

Retiring in 1973, he resumed individual practice as a consulting engineering geologist and is currently active in his practice. He was elected to the National Academy of Engineering in 1987. He and Winifred, his wife of fifty years, currently live in Vallejo, California. They have three children; Alan, a professor and researcher at Carnegie-Mellon University and Teri and Diana who reside in California. They also have six grandchildren.

## 1962-63 CARTER

E. H. CARTER was born October 3, 1904 in Leadville, Colorado. After graduation from the University of Colorado in 1926, with a B.S. in electrical engineering, Carter embarked on a long and varied engineering career on projects throughout the western United States; Colombia, Venezuela, Brazil, Peru and Ecuador in South America; Pakistan and the Dominican Republic. For a period of 7 years he was consultant to the Venezuelan Government on 26 projects including dams, irrigation systems, hydro-electric projects, drainage and river control.

He was with R.J. Tipton & Associates in Denver and a partner in the firm of Phillips-Carter-Osborn. From 1967-1969 he was a member of the Regional Export Expansion Council, appointed by the U.S. Secretary of Commerce.

Prior to his death, Carter was active in several technical societies; a Fellow of the American Society of Civil Engineers, a member of Consulting Engineers Council of Colorado, Colorado Society of Engineers, Wyoming Society of Engineers, Professional Engineers of Colorado, Colorado Engineering Council and Colorado Construction League. He and his wife had three children.

## 1963-64 KONKEL

EDWARD VERNON KONKEL was born in LaJunta, Colorado on April 28, 1923 and was called "Vern." He received his bachelor's degree in Architectural Engineering from the University of Colorado in 1948 after serving as a pilot in the U.S. Navy from 1943 to 1945. Upon graduation he began his engineering career as a design engineer in the office of Milo S. Ketchum in Denver. He received a M.S. in civil engineering in 1955 from the University of Colorado.

In 1954 he entered into partnership with Ketchum and the firm name became Ketchum and Konkel, Consulting Engineers. During the ensuing years the name of the firm changed as other partners were taken into the firm. In October, 1968 the firm name became Ketchum Konkel Barrett Nickel Austin. Konkel served the firm as president from 1963 to 1969 and as chairman of the Board of Directors until his death on Christmas day, 1970. One of the many notable structures which Konkel personally designed was the Broadmoor Hotel International Center, Colorado Springs, which is reputedly the largest hyperbolic paraboloidal shell of its kind in the world.

A fellow of the American Society of Civil Engineers, he was an active member of the International Association of Shell Structures, Professional Engineers of Colorado (president 1958), Structural Engineers Association of Colorado (president 1959), CECC and ACEC. He was a U.S. delegate from the Council to the International Federation of Consulting Engineers in Helsinki, 1962; London, 1963; and Paris, 1964. He was a major force in the establishment of a Colorado Council of Professions in Denver in 1967. The coalition was composed of representatives from the Bar Association, Dental Association, Medical Association, Society of Architects, Society of Certified Public Accountants, and the Consulting Engineers Council. The group provided coordination between professions.

Konkel authored several papers and articles for various professional magazines. He was awarded the professional degree of Civil Engineer at the University of Colorado in 1960. In 1966, he was named "Colorado Engineer of the Year" and in 1972 was the recipient (posthumously) of a Distinguished Engineering Alumnus award from the University of Colorado College of Engineering. Vern married Rojean Joy Templeman in Oregon, Illinois in 1949. Six children were born of this union: R. Steven, Susan, Gregory, Jeffrey, Michael and Mary.

## 1964-65 BUNTS

JOHN E. BUNTS is a native of Colorado, born in Colorado Springs. He received his Bachelor of Science degree in Architectural Engineering from the University of Colorado. Bunts' career began with Phillips, Carter, Osborn, Denver, Colorado as a designer. He subsequently became a partner in the firm, Bunts, Kelsey and Bunts, Architects in Colorado Springs. He then began his own consulting engineering firm, Jack Bunts, Consulting Engineer in Colorado Springs, and later he joined the firm of Charles T. Main, Denver, Colorado as their branch manager.

He eventually moved to Phoenix, Arizona where he was division head for the architectural department of Johannessen and Girand and is currently providing consulting engineering services to this firm. A major design project engineered by Bunts was Moby Gym at Colorado State University in Fort Collins, Colorado. The gym was named "Moby" by the CSU students as part of a contest because of its large size, whale-line shape and appearance. John, the 10th president of CECC, is the father of three daughters and one son.  He has four grandchildren.

## 1965-66 MEURER

CHARLES S. MEURER, a southern Indiana farm boy, was born in 1920.  Instilled with a great love of sports, he became an athlete as well as student at Rose-Hulman Institute where he graduated as a Civil Engineer.  In 1955 he entered the Colorado professional engineers life as one of the founders of a firm, Meurer-Serafini-Meurer, Inc., and over the next twenty-one years became very involved with the growth of the community and the neighboring states.  The firm's experience covered all phases of engineering: design and surveying on dams, airports, bridges, water supply, sewage disposal, highways and land development.

In 1976 on completion of the sale of the firm in Colorado, Charles moved to California to join the firm of Meurer Engineering, Inc. (now in Santa Clarita) started earlier by his son, Roger.  Joining them was another engineering son, James.  The middle son, Charles Lonnie has a manufacturing firm, TRAC-VAC, in Denver which uses some of his own patents.  Charles and Helen, his wife, now have three grandsons and a granddaughter.

## 1966-67 CLEVENGER

WILLIAM A. CLEVENGER was born in Wheatland, Wyoming in 1919, and lived in Sunrise, Wyoming until 1937.  He attended the University of Wyoming, receiving a Bachelor of Science degree in Civil Engineering in 1943.  After serving in the U.S. Army for 3-1/2 years, he married and moved to Denver, Colorado where he worked for the U.S. Bureau of Reclamation for 10 years.  He went with Woodward-Clyde Consultants in 1956 in charge of their Denver practice for 17 years, ending up as Chairman of the Board until he retired in 1984.

Clevenger has been quite active in professional societies.  He has been president of the Colorado sections of the Consulting Engineers Council of Colorado, and of the American Society of Civil Engineers.  He also served as vice president and president of the American Consulting Engineers Council and served a six year term on the Executive Committee of the U.S. Committee on large dams.

He has published many papers and has co-authored a book, "Earth and Earth Rock Dams," John Wiley & Sons, 1963.  Bill and his wife, Janet,  have 4 children, 11 grandchildren, and currently live in Sequim, Washington.

## 1967-68 KONKEL

JAMES (Jim) KONKEL was born in LaJunta, Colorado in 1925 and finished high school there. He entered the U.S. Navy V-12 program and was commissioned as Ensign, USNR in 1945. Jim graduated from the University of Colorado, Boulder in 1949 with a bachelor's degree in Chemical Engineering. After several positions he opened an office in Denver, practicing as a consulting engineer in mechanical system designs for buildings. The firm is now McFall, Konkel & Kimball. Konkel has been president of the Consulting Mechanical Engineers Association of Colorado as well as Consulting Engineers Council of Colorado.

Konkel has received the ASHRAE Fellow Award and the ASHRAE Energy Conservation Award as well as the University of Colorado Distinguished Engineering Alumnus Award for Engineers in Private Practice, 1981. He is a fellow in the American Consulting Engineers Council. He received the Orley O. Phillips Award from CECC in 1983. He is the author of "Rule of Thumb Cost Estimating for Mechanical Systems in Buildings." Jim and his wife, Joyce, have two sons. One is a consulting engineer. They have three grandchildren.

## 1968-69 HAWES

WILLIAM H. HAWES was born in Tyler, Texas on September 22, 1928. He moved to Colorado in 1951. He has a B.S. in Civil Engineering from Texas A & M, and a M.S. in Civil Engineering from the University of Colorado. He worked for the American Bridge Company in Gary, Indiana; Milo Ketchum in Denver; and Rhuel Andersen, eventually becoming a partner in Andersen, Koerwitz & Hawes. He eventually formed his own company, Hawes & Associates. His favorite project was the Denver Art Museum.

He was General Chairman of the ASCE Annual Convention, 1975. He is the father of three children: Joan Cunningham, Thomas, and Andrew, all living in Denver. His writings include, "Plastic Design of Continuous Floor Beams," AISC Engineering Journal.

## 1969-70 WRIGHT

KENNETH R. WRIGHT was born in Milwaukee, Wisconsin in 1929 and received his Bachelor of Science and Master of Science degrees in Civil Engineering from the University of Wisconsin. He also holds a Bachelor of Business Administration degree. His early experience was with Arabian American Oil Company in Saudi Arabia, the U.S. Bureau of Reclamation and W.W. Wheeler and Associates.

KEN WRIGHT Continued

He moved to Colorado in 1957 and founded his own firm, Wright Water Engineers in 1960. A significant Colorado project Ken worked on is the Adolph Coors Company's water supply system. He is the recipient of the U.S. Department of Interior's Valor award. Wright is married and has two daughters. His wife, Ruth, is an attorney and a Colorado State Legislator.

## 1970-71 BECKER

RALPH W. (Bill) BECKER, born in Buffalo, New York, and attended elementary and high school in Detroit, Michigan. He became an Eagle Scout and competed in track. He received a B.S. in Civil Engineering from Michigan State University in 1943. He married Elaine J. Wagar in 1936. After three years in plant engineering for General Motors, he spent a year and a half with the U.S. government C.A.A. locating and installing airways facilities in the western states. Returning to Detroit he joined the Albert Kahn Company and worked on designs for U.S. Naval Air Stations.

Becker moved to Denver in December 1939 and took a design assignment with the Bureau of Reclamation. This work included the designs for the 11 span concrete arch bridge over Grand Coulee Dam.

He founded Technical Service Company (Tesco) in 1947. In the early 50's, Tesco began serving the newspaper publishing industry. Complete industrial processing and plant design services were provided for over 100 major publishing operations. A significant project was the new Detroit News Sterling Heights Plant wherein the company did all the planning, systems design, equipment procurement, plant design and the management of construction and installation.

Technical Service Company was sold to Charles T. Main of Boston in 1972. Becker stayed on with the operation until he retired in 1975. Elaine and Bill have two daughters, Susan Bently of Aurora and Barbara Franz of Maryland; also two grandsons, Thomas and Matthew Franz.

## 1971-72 FLEMING

DAVID E. FLEMING, a native Coloradoan, received his B.S. degree (with special honors) and was canebearer of the 1943 graduating class of the University of Colorado. In 1951 he was awarded the M.S. degree from the same institution. After graduate work at the University of Chicago, he served as a meteorology officer in the U.S. Army Air Corps during World War II.

## DAVID FLEMING Continued

A design and supervisory engineer with the U.S. Bureau of Reclamation, Denver, from 1946 to 1957, he then served as a principal in the firms of A.J. Ryan and Associates; Ketchum, Konkel, Ryan and Fleming; The David E. Fleming Company and Ketchum Konkel Barrett Nickel Austin until his retirement in March, 1986.

He is a past-president of the Rotary Club of Denver, Consulting Engineers Council/ Colorado and Professional Engineers of Colorado; a life member of the American Society of Civil Engineers and the National Society of Professional Engineers; a fellow in the American Consulting Engineers Council; a member of the Colorado Society of Engineers; a Certified Consulting Engineer, Colorado; and a retired Professional Engineer, Colorado. He received a Distinguished Engineering Alumnus Award from the University of Colorado College of Engineering in 1973. Dave and his wife, Lois, have two daughters and one grandson.

## 1972-73 MCFALL

JAMES (Jim) D. MCFALL was born in Sabetha, Kansas in 1935. His family moved to Denver when he started grade school. He attended Denver public schools and the University of Denver where he obtained a B.S. in Chemical Engineering. He served as an officer in the U.S. Army Chemical Corp for two years. In 1960 he returned to Denver to work for James H. Konkel.

In 1965, Jim Konkel and Jim McFall formed a partnership of McFall & Konkel Consulting Engineers which later became McFall, Konkel & Kimball Consulting Engineers. To date these firms have designed over 5,000 mechanical and electrical systems for building projects. A significant project designed by the firm is the Engineering School at the University of Colorado, Boulder.

McFall has been a member of CECC since 1965 and since then has attended all but one of CECC's annual state conferences. He missed one when his daughter was married on the same weekend as the conference. Jim and his wife, Margaret (Marge), have three children; Barbara, Mary, and Matthew.

## 1973-72 STEWART

JAMES H. STEWART was born in Hanford, California in 1923 and received his Civil Engineering degree from the University of California at Berkeley in 1944. During World War II, he served as a Naval officer in the Pacific Theater. From 1946 through 1956, he was employed as a construction engineer and project manager by several major contractors on heavy construction projects throughout the western states.

In 1957 he formed his own consulting firm, James H. Stewart & Associates, Inc. in Fort Collins, Colorado for the practice of general civil engineering and land surveying. He also founded, in 1960, Empire Laboratories, a geotechnical and materials testing firm. He served as president of both firms until the early 1980's when he sold his interests to long time associates. In 1962, he was elected to the office of Larimer County Surveyor and served in this office for six terms until 1986 in addition to his private practices.

In 1985 he and his son David R. Stewart, founded Stewart Environmental Consultants, Inc. for the practice of environmental engineering. He and his wife, Fran, have four children and six grandchildren and maintain homes in Fort Collins and Palm Desert, California.

**1974-75 CHEN**

FU HUA CHEN came to Colorado in 1957 and has been here ever since. He was born in Shanghai, China, July 21, 1912. He received his B.S. Civil Engineering from the University of Michigan in 1935 and his M.S. Civil Engineering from the University of Illinois. In 1980, he received an Honorary Doctorate of Science from Colorado State University.

Chen is founder of Chen & Associates, a geotechnical consulting firm. Among their many projects are the Kodak Manufacturing Plant in Windsor; the Mile High Stadium; Grayrocks Dam in Wheatland, Wyoming and a significant number of high-rise structures in Denver. Chen is a world-renowned expert in the area of expansive soils and has authored numerous articles and a text book on the subject, as well as conducting seminars throughout the world.

From 1949 through 1956, Chen was responsible for supervising all materials testing and foundation investigations for the Hong Kong Public Works Department. Earlier, he held a professorship at National Fu Tan University in Shanghai, where he taught soil mechanics and highway engineering. As chief engineer for China's Ministry of Communications from 1940 to 1945, he was responsible for the construction of three major highways: the 600-mile-long Koknor-Tibet Highway, the 717-mile-long Burma Road and the 400-mile-long Szechuan-Sikang Highway.

Chen is active in several professional societies. He is a fellow of the American Society of Civil Engineers, has served as president of the Colorado section of ASCE, and chaired the ASCE Geotechnical Division. He also served as president of the Professional Engineers of Colorado. His activities with the Colorado State Board of Registration for Professional Engineers and Land Surveyors include a term as Vice Chairman. He spends much of his time on legislative and education matters on behalf of various engineering societies. He has been a panelist for the American Arbitration Association for ten years. He has lectured extensively at Colorado State University, has served on faculty and advisory committees.

Fu Hua has received numerous honors including the 1984 Civil Engineer of the Year Award by the Colorado Section of ASCE, the 1983 Personal Service Award from Colorado State University, the CECC Orley O. Phillips Award in 1982, the Colorado Engineering Council Gold Medal Award in 1982, the 1981 Professional Engineers of Colorado Alfred Ryan Award and an award for Outstanding Service from the American Society of Civil Engineers in 1978. In 1944 Chiang Kai-Shek personally presented him with the National Award for the completion of the Burma Road. Fu Hua and his wife, Edna, have three children and four grandchildren.

## 1975-76 RICE

LEONARD RICE was born in New London, Connecticut in 1932. He received his B.S. in Civil Engineering from Auburn University in 1954 and M.S. in Civil Engineering from the University of Iowa in 1961. Following two years of duty as a Navy officer he spent four years with a consulting firm in Florida then, after a year in graduate school, he moved to Denver in 1961. He was with Tipton and Kalmbach from 1961 to 1968 and Wright Water Engineers from 1968 to 1970.

In 1970 he formed Leonard Rice Consulting Water Engineers, Inc., (LRCWE) specializing in water resource development and water rights. LRCWE received a national ACEC Engineering Excellence award for their Boulder White Rocks project in 1974.

Rice has taught water resource engineering at Denver University and the University of Colorado, Denver and is co-author of "Engineering Aspects of Water Law." In 1987 he received the prestigious CECC Orley O. Phillips award. For Lee and his wife, Kay, the highlight of their CECC association was serving as an ACEC vice president for 1985-87, which involved meeting and working with ACEC members throughout the USA. He is the father of three children and has four grandchildren.

## 1976-77 ANDERSON

ALBERT E. ANDERSON was born in Denver, Colorado in 1932. He was educated in Denver public schools and graduated from the University of Colorado in 1957 after serving two years in the U.S. Army Combat Engineers during the Korean War.

Upon graduation, Anderson went to work as a consulting structural engineer with the Denver firm of Ketchum and Konkel. In 1968, Jim and Al formed their existing firm of Anderson and Hastings, Consulting Engineers. That firm has completed over 4,000 projects since its conception.

During his term as president of the council, Al actively and successfully pursued legislation dealing with frivolous and third party suits that were causing extensive costs to consulting engineers. He and his wife, Alice, have two children, Stewart and Scott.

## 1977-78 BLATCHLEY

RONALD K. BLATCHLEY was born near Kearney, Nebraska in 1929. His father moved the family to an irrigated farm near Fort Collins, Colorado in 1939, where Ron subsequently attended Colorado State University, graduating in 1951 with a B.S. in Civil Engineering.

After serving as an officer in the U.S. Air Force during the Korean War, he worked for the water resource consulting firms, Tipton and Kalmbach, Inc. and W. W. Wheeler and Associates, working on such projects as Denver Water Board's Dillon Dam and Roberts Tunnel and the development of West Pakistan's Indus Basin water resource projects in addition to attending the University of Colorado graduate school.

In 1969 he organized the firm Blatchley Associates specializing in water rights engineering for such agencies as Englewood, Vail, Thornton, Evergreen, Jefferson County School District R-1, as well as private industry firms such as Brannan Sand and Gravel, Mobile Premix Concrete, Inc., Hartford Insurance Company and Phillips Petroleum Company. Significant projects include the water supply for Meridian Office Park, Mahogany Oil Shale project and water rights for many municipal, industrial and residential developments.

While CECC president, Blatchley was the organizer and first chairman of Consulting Engineers Political Action Committee. He also served as chairman of the ASCE's Irrigation and Drainage Division and received the Division's "Outstanding Service Award" in 1988. Ron and his wife, Alvena, have lived in the Denver Metro area since their marriage in 1951. To date, their four daughters have provided them with nine grandchildren.

## 1978-79 NICKEL

DONAVON D. NICKEL was born in Bismarck, North Dakota in 1929. He attended North Dakota State College where he received a degree in Architectural

Engineering in 1952. He attended the University of Illinois and received a master's degree in Civil Engineering in 1953. He served as a First Lieutenant in the U.S. Air Force until 1956. In 1963, because of his interest in management, Nickel obtained a master's degree in Business Administration.

From 1956 to 1987, he was a consulting engineer practicing structural engineering. He was a principal and partner in the firm of Ketchum Konkel Barrett Nickel Austin, Inc. Long active in consulting engineering, he served as the 24th president of Consulting Engineers Council of Colorado. He has three sons and one granddaughter.

## 1979-80 KOONSMAN

GEORGE L. KOONSMAN, the 25th president of CECC, was born in Lamar, Colorado in 1922. He received his Bachelor of Science degree at Colorado State University in 1947 and his Master of Science degree from the same university in 1950. He was an instructor at Colorado State University from 1947 to 1951.

Koonsman went to work for Ideal Cement Company as a project manager in 1951. From 1959 to 1962 he worked for Martin Marietta (Cement) as a cement engineer. He then worked for the Ken R. White Company as senior vice president from 1962 to 1974. A major project in which Koonsman was responsible was the design of the Seattle Plant for Ideal Cement Company which won the CECC Grand Award for Engineering Excellence in 1967.

George was with Centennial Engineering from 1974 to 1986 where he served as president and chairman of the board. He is a life member of CECC, a fellow of ACEC, a fellow of ASCE and an Honor Alumnus of Colorado State University. George is survived by his wife, Virginia, and three children; Carl, Janet, Thomas.

## 1980-81 BISHOP

HAROLD F. BISHOP was born in Long Beach, California in 1935. After graduation from the University of Utah in 1958 with a Bachelor of Science in Civil Engineering, he served three years with the U.S. Navy as engineering officer aboard a destroyer. Following release from active duty, he was with the U.S. Forest Service in Ogden, Utah for two years before moving to Denver and joining the firm of Tipton and Kalmbach of Denver, Colorado. After one year in that firm's Denver office, he was transferred to Pakistan where he served in various assignments until 1970.

Bishop returned to Denver and was a principal with two Denver area consulting firms between 1970 and 1980 prior to forming Bishop Associates, Inc. He has been active in numerous civic organizations. In addition to being president of the Consulting Engineers Council of Colorado, he was the association's national director from 1986 to 1988. He and his wife, Ila, have two children and two grandchildren.

## 1981-82 HEPWORTH

RICHARD C. HEPWORTH was born in Topeka, Kansas in 1930. He graduated from high school in Tulsa, Oklahoma in 1948 and entered Rice University. In 1952, he joined the Air Force and went through pilot training and was discharged in 1956 with the rank of a First Lieutenant. He later participated in reserve activities reaching the rank of Major. He completed his schooling at the University of Colorado, graduating with a B.A. degree in Geology in 1959. The next year he worked as a physical oceanographer for the Navy in Washington, D.C., participating in ocean surveys for anti-submarine warfare. He next attended the University of Utah, graduating with a M.S. degree in Geological and Civil Engineering in 1963.

From 1960 to 1965, he worked for the Utah Highway Department, rising to head of the foundations department. In 1965, he moved to Denver as a senior project engineer with Woodward-Clyde-Sherard & Associates. In 1967, he accepted a position with Leedshill-DeLeuw in Bangladesh working on the coastal embankment project.

Hepworth joined Chen & Associates in 1969 and rose to president of the firm in 1984, a position he held until 1988. He has participated in many major building projects, high-rise structures, dams and mining facilities throughout the western United States. He is active in professional organizations and is a past president of Colorado Engineering Council as well as CECC. Dick has three daughters.

## 1982-83 BARRETT

MICHAEL H. BARRETT. A native of Colorado, he was born in Dove Creek, June 20, 1932 and educated in the state. He holds a Bachelor of Science degree in Civil

Engineering from the University of Colorado and in 1979 received a MBA from the University of Denver. In his 33 years with KKBNA he has held several positions including project manager, associate, chief engineer, and principal.

He was president and/or chairman of KKBNA during a period of growth from a one office operation to expansion into offices in Glenwood Springs, Colorado Springs, Salt Lake City, New York, Chicago, Boston, St. Louis, Bellingham (WA) and Vancouver (BC), as well as Denver; providing design services in 40 states. His favorite project still remains Currigan Convention Center with its state-of-the art steel space frame roof. Currently he is principal consultant for Martin/Martin Consulting Engineers.

Active in many community as well as professional organizations, Mike has served as president of the Denver area Boy Scouts of America and Professional Engineers of Colorado, as well as CECC. Barrett is the recipient of several honors including a Distinguished Engineer award from the University of Colorado College of Engineers in 1984.

He holds patents for a "linear sphere" space frame roof connection and for "diadeck," a structural floor system. His wife is Barbara Krentz Barrett and they have two sons, two daughters and two grandsons. His oldest son is a consulting engineer.

## 1983-84 HASTINGS

JAMES M. HASTINGS was born on a farm near Longmont, Colorado in 1923. His parents moved to Sterling, Colorado where he attended schools and graduated from Sterling High in 1939. He joined the U.S. Navy in 1942 and served as a First Lieutenant on a destroyer in the Pacific Theatre. He graduated from the University of Colorado at Boulder as a Civil Engineer in 1948.

His engineering experience includes Stanolind Oil & Gas Company offshore development; Bureau of Reclamation; Silas Mason & Company; Stearns-Roger, Ketchum, Konkel, Ryan and Hastings; and as a principal of Anderson & Hastings for the past twenty years. A significant project he provided design services on is Thomas Jefferson High School. Jim and his wife Della, have three children; Judith, Donna and Jim V., and two grandchildren.

## 1984-85 DUTZI

HOWARD C. DUTZI, the 30th president of CECC, was born in Chicago, Illinois in 1930. He received his Bachelor of Science degree at the University of Illinois in 1953

and his Master of Science degree from the same university in 1955. He worked for Skidsmore, Owings & Merrill, Chicago, from 1955-58, fifteen months of which was spent in Colorado Springs as construction phase structural engineer for various buildings at the Air Force Academy.

He moved to Colorado Springs in 1958 and worked for a local architectural firm until 1961 when he organized his own consulting structural engineering firm. His firm performed the structural engineering design in Colorado Springs for such projects as the Pikes Peak Center, First National Bank Building, United Bank Building, El Paso County Jail, and the Airport Terminal Building. He has also done structural engineering on building types throughout Colorado and elsewhere in the nation.

Dutzi has been involved in many civil and professional organizations over the past 25 years, including being president of the Pikes Peak Chapter of the Construction Specifications Institute and the Colorado Springs Executive Association. He is currently the president of the Colorado Springs Symphony Council. Howard is married to Ruth, who is a junior high school principal.

## 1985-86 MULLER

LARRY A. MULLER was born in Mankato, Minnesota in 1943. Through the influence of his father, a producer of limestone rock products, he enrolled in a pre-engineering curriculum in college and ultimately earned a Bachelor and a Master of Science degree in Civil Engineering at the University of Minnesota. During his college years, he worked summers for the Minnesota Department of Highways in the Student/Trainee program. He later worked on hydraulic engineering research projects at the St. Anthony's Falls Hydraulic Laboratory on the Mississippi River at Minneapolis. Aside from his military tour of duty at the Cold Regions Research and Engineering Laboratory in Hanover, New Hampshire, his total professional career has been as a consulting engineer.

He joined a consulting firm in Denver in 1971 and immediately became involved with planning and design engineering for one of the first major drainageway improvement projects for the Denver area Urban Drainage and Flood Control District. Following the disastrous Big Thompson River flood of 1976, he led the engineering effort to define the areas along the river for restricted redevelopment for the Larimer County Commissioners.

He formed Muller Engineering Company, Inc. in 1980. A significant recent project completed by his firm is the remodeling and expansion of Crossroads Mall. Larry and his wife, Ellen, have a son and daughter.

## 1986-87 HOLT

STEPHEN A. HOLT was born in Horace, Kansas in 1935. He received his educational training in Colorado schools, graduating from the University of Colorado in 1958 with a Bachelor of Science in Civil Engineering. He worked for the Bureau of Public Roads, where he designed and provided construction management of highways in national parks and forest until 1965, at which time he went to Ecuador, South America for TAMS to work on a 385 kilometer highway system traversing the Andes.

In 1967, Holt worked for the Wisconsin DOT as a design squad leader and returned to Colorado in 1970. His career continued with NHPQ Consultants and was involved in transportation facilities, sewage treatment projects, and land development assignments. He served as president of URS Engineers, Inc. in the Rocky Mountain area and was involved in numerous large multi-discipline civil engineering projects.

In 1985 he formed his own firm of Felsburg Holt & Ullevig, specializing in transportation engineering. A significant project his firm is providing services on is the master planning for the access to the proposed new Denver airport. Steve has a wife, Louise, one son, a daughter and one grandchild.

## 1987-88 EMMONS

ROBERT E. EMMONS was born in East Liverpool, Ohio in 1934. He received his Bachelor of Science degree in Civil Engineering at Ohio University. He began his career with Madison Engineering Associates in Mansfield, Ohio as a design engineer. He then went to work for G.W. Raite and Associates in Ashland, Ohio as chief engineer. Emmons next located in Tuscon, Arizona where he became chief engineer of Anderson Engineering Company and then moved to Decatur, Illinois to become associate and branch manager of Warren and Van Pragg, Inc.

He came to Denver, Colorado in 1972. He was president of EHMG, Inc., consulting engineers prior to founding R & R Engineers-Surveyors, Inc. in 1988. He is currently president of the firm. He is past president of the Economic Development Group (ECHO) in Aurora, Colorado. He received the Man-of-the-Year award in 1976 from the Aurora Chamber of Commerce as well as a Distinguished Service Award from the Junior Chamber Commerce, Ashland, Ohio. Bob and his wife, Eula, have three children, Suzan, Sharon, and Robert Jr., and two grandchildren.

## 1988-89 WEINGARDT

RICHARD WEINGARDT. A third generation native of Colorado, and the son of a general contractor, Weingardt was born and raised in Sterling, Colorado where he graduated as valedictorian from St. Anthony's High School. He received both his Master of Science and Bachelor of Science degrees in Civil Engineering from the University of Colorado. His engineering career began in 1960 with the United States Bureau of Reclamation in California and Colorado. He founded his own firm, Richard Weingardt Consultants, Inc., in 1966 after working for Ketchum Konkel Ryan and Fleming.

Licensed in 29 states, he has directed his firm's projects in 40 states and several foreign countries. His firm's United Airlines Concourse B at Stapleton International Airport project is the recipient of the 1988 CECC Grand Award and the ASCE/ Colorado Civil Engineering Project-of-the-Year honors.

Active in community as well as professional organizations, Weingardt is the recipient of several honors including: the 1987 Outstanding Alumnus Award from the University of Colorado, a Distinguished Engineer Award from the University of Colorado College of Engineering in 1988, and the 1987 SMPS Leonardo Award. He was appointed by Governor Romer to the State Historic Preservation Review Board in 1987 and to "Vision Colorado" by the Colorado State Legislature in 1988.

Weingardt is the author of numerous technical papers and two books, "Sound the Charge" and "Circular Cylindrical Shell Design." He lectures to businesses, students and at conferences in this country as well as abroad on engineering, leadership and business. Rich and his wife, Evelyn, are the parents of three children, Nancy, Susan and David.

*The 35th president of CECC will be William Moore whose term will run from spring 1989 to spring 1990.*

*WILLIAM W. MOORE was born in Pasadena, California. He received his B.S. and M.S. in Civil Engineering at Stanford University. He also received a MBA in 1967 at the University of California (Berkeley). His career includes experience with Dames and Moore. He is currently general manager of International Technology Corporation. Bill and his wife, Sandy, have two children.*

# Roster of Officers

**1955 CECC OFFICERS:**

President, Ken Murray
Vice President, Alvin D. Swanson
Secretary-Treasurer, Robert Behrent

**1956 CECC OFFICERS:**

President, Milo Ketchum
Vice President, Kenneth White
Secretary-Treasurer, Ib Jorgensen

**1957 CECC OFFICERS:**

President, Rhuel Andersen
Vice President, Robert Behrent
Secretary-Treasurer, Albert McKittrick
National Director, Orley O. Phillips

**1958 CECC OFFICERS:**

President, Paul Koch
Vice President, Vernon Konkel/R. Behrent
Secretary-Treasurer, John E. Tracy
National Director, Orley O. Phillips

**1959 CECC OFFICERS:**

President, Alvin Swanson
Vice President, Francis Stark
Secretary-Treasurer, Eugene B. Waggoner
National Director, Orley O. Phillips

**1960 CECC OFFICERS:**

President, John E. Tracy
Vice President, William A. Clevenger
Secretary-Treasurer, Jan A. Norris
National Director, Vernon Konkel

**1961 CECC OFFICERS:**

President, Eugene B. Waggoner
Vice President, John E. Bunts
Secretary-Treasurer, Kenneth R. Wright
National Director, E. Vernon Konkel

**1962 CECC OFFICERS:**

President, E.H. Carter
Vice President, C. Kenneth Kolstad
Secretary-Treasurer, Clifford Johnson
National Director, E. Vernon Konkel

**1963 CECC OFFICERS:**

President, E. Vernon Konkel
Vice President, James M. Hastings
Secretary-Treasurer, H. Joe Meheen/James Konkel
National Director, Charles Meurer

**1964 CECC OFFICERS:**

President, John E. Bunts
Vice President, James H. Konkel
Secretary-Treasurer, Vernon S. Winkel
National Director, Rhuel A. Andersen

**1965 CECC OFFICERS:**

President, Charles S. Meurer
Vice President, Kenneth R. Wright
President-Elect, William A. Clevenger
Secretary-Treasurer, Fred S. Rink
National Director, Rhuel A. Andersen
Directors:  John E. Bunts, William H.
Hawes, Walter Langebartel, Eugene
Waggoner, Sol Flax and Malcolm
Meurer

**1966 CECC OFFICERS:**

President, Bill Clevenger
Vice President, Walter Langebartel
Secretary-Treasurer, Bob Rice
National Director, Kenneth R. Wright
Directors:  (No records available)

**1967 CECC OFFICERS:**

President, James Konkel
Vice President, William Hinch
Secretary-Treasurer, Bill Bredar
National Director, Kenneth R. Wright
Directors:  (No records available)

**1968 CECC OFFICERS:**

President, William Hawes
Vice President, Fred Rink
Secretary-Treasurer, Sol Flax
National Director, William Clevenger

**1969 CECC OFFICERS:**

President, Kenneth R. Wright
Vice President, Sol Flax
Secretary-Treasurer, James D. McFall
President-Elect, Bill Becker
Past President, William H. Hawes
National Director, William Clevenger
Directors:  Bob Voiland, Bob Nedell,
Olin Kalmbach, Orley Phillips, James
Stewart, George Williams and Malcolm
Meurer
*Executive Director, Harvey A. Kadish*
104

**1970 CECC OFFICERS:**

President, Ralph W. (Bill) Becker
Vice President, James D. McFall
Secretary-Treasurer, Fu Hua Chen
President-Elect, David E. Fleming
Past President, Kenneth R. Wright
National Director, Alvin Swanson
Directors:  Orley Phillips, George
Williams, James Stewart, Mike Barrett,
George Koonsman, C. Kenneth Kolstad
and Malcolm Meurer
*Executive Director, Harvey (Hak)
Kadish*

**1971 CECC OFFICERS:**

President, David E. Fleming
Vice President, Fu Hua Chen
Secretary-Treasurer, James M. Hastings
President-Elect, James D. McFall
Past President, Ralph W. (Bill) Becker
National Director, Alvin Swanson
Directors:  C. Kenneth Kolstad, Mike
Barrett, George Koonsman, Russell
Miller, Donavon Nickel, Stanley
Thorfinnson, Malcolm Meurer
*Executive Director, Harvey (Hak)
Kadish*

**1972 CECC OFFICERS:**

President, James D. McFall
Vice President, Mike Barrett
Secretary-Treasurer, Edward Bierbach
President-Elect, James Stewart
Past President, David E. Fleming
National Director, Fu Hua Chen
Directors:  Russell Miller, Donavon
Nickel, Stanley Thorfinnson, Andrew
Pfeiffenberger, LaVern Nelson
*Executive Director, Hak Kadish*

**1973 CECC OFFICERS:**

President, James H. Stewart
Vice President, Albert E. Anderson
Secretary-Treasurer, Leonard Rice
President-Elect, Fu Hua Chen
Past President, James D. McFall
National Director, Fu Hua Chen
Directors:  Andrew R. Pfeiffenberger,
Charles E. Bell, Gerald V. Kohnert,
Howard C. Dutzi, LaVern C. Nelson,
Gordon A. McDonald

**1974 CECC OFFICERS:**

President, Fu Hua Chen
Vice President, Stanley Thorfinnson
Secretary-Treasurer, Robert S. Nedell
President-Elect, Leonard Rice
Past President, James H. Stewart
National Director, David E. Fleming
Directors:  Charles E. Bell, Gordon A.
McDonald, Howard C. Dutzi, Ronald K.
Blatchley, Richard C. Hepworth, Russell
M. Miller
*Executive Director, Edward W. Thorson*

**1975 CECC OFFICERS:**

President, Leonard Rice
Vice President, William L. Bredar
Secretary-Treasurer, Charles E. Bell
President-Elect, Albert E. Anderson
Past President, Fu Hua Chen
National Director, David E. Fleming
Directors:  Ronald K. Blatchley, Richard
C. Hepworth, Russell M. Miller,
Kenneth C. Kolstad, H. Joe Meheen,
Charles W. Meyers
*Executive Director, Edward W. Thorson*

**1976 CECC OFFICERS:**

President, Albert E. Anderson
Vice President, Edward R. Bierbach
Secretary-Treasurer, Donavon D. Nickel
President-Elect, Ronald K. Blatchley
Past President, Leonard Rice
National Director, William H. Hawes
Directors:  C. Kenneth Kolstad, Kenneth
D. Bielman, H. Joe Meheen, Dwight R.
Sayles, Charles W. Meyers, Chester C.
Smith
*Executive Director, Edward W. Thorson*

**1977 CECC OFFICERS:**

President, Ronald K. Blatchley
Vice President, George L. Koonsman
Secretary-Treasurer, Richard C.
Hepworth
President-Elect, Donavon D. Nickel
Past President, Albert E. Anderson
National Director, William H. Hawes
Directors:  Kenneth D. Bielman, R.
Keith Hook, Dwight R. Sayles, Harold
F. Bishop, Chester C. Smith, Frank J.
Holliday
*Executive Director, Ed Thorson*

**1978 CECC OFFICERS:**

President, Donavon D. Nickel
Vice President, Chester C. Smith
Secretary-Treasurer, Dwight R. Sayles
President-Elect, George L. Koonsman
Past President, Ronald K. Blatchley
National Director, Leonard Rice
Directors:  H. Jack Kraettli, Wayne C.
Irelan, Frank J. Holliday, Robert C.
McWhinnie, Harold F. Bishop, Robert
W. Thompson
*Executive Director, Ed Thorson*

**1979 CECC OFFICERS:**

President, George L. Koonsman
Vice President, Richard C. Hepworth
Secretary-Treasurer, David E. Austin
President-Elect, Harold F. Bishop
Past President, Donavon D. Nickel
National Director, Leonard Rice
Directors:  Wayne C. Irelan, Robert C.
McWhinnie, Robert W. Thompson,
Leonard C. Becker, James M. Hastings,
Larry A. Muller
*Executive Director, Ed Thorson*

**1980 CECC OFFICERS:**

President, Harold F. Bishop
Vice President, Charles E. Meyers
Secretary-Treasurer, Wayne C. Irelan
President-Elect, Richard C. Hepworth
Past President, George L. Koonsman
National Director, Albert E. Anderson
Directors:  Leonard C. Becker, James M.
Hastings, Larry A. Muller, Victor H.
Weidmann, Deryl W. Gingery, Charles
J. Reich
*Executive Director, Edward Thorson*

**1981 CECC OFFICERS:**

President, Richard C. Hepworth
Vice President, Wayne C. Irelan
Secretary-Treasurer, Larry Muller
President-Elect, Michael H. Barrett
Past President, Harold F. Bishop
National Director, Albert E. Anderson
Directors:  Victor H. Weidmann, Oliver
Watts, Deryl W. Gingery, Gordon C.
Meurer, Charles J. Reich, George D.
Sellards
*Executive Director, Sondra S. Smith*

**1982 CECC OFFICERS:**

President, Michael H. Barrett
Vice President, Howard C. Dutzi
Secretary-Treasurer, Charles J. Reich
President-Elect, James M. Hastings
Past President, Richard C. Hepworth
National Director, James Konkel
Directors:  Gordon C. Meurer, Stephen
A. Holt, George D. Sellards, W. Tom
Pitts, Oliver E. Watts, Richard
Weingardt
*Executive Director, Sondra S. Smith*

**1983 CECC OFFICERS:**

President, James M. Hastings
Vice President, David E. Austin
Secretary-Treasurer, Gordon C. Meurer
President-Elect, Howard C. Dutzi
Past President, Michael H. Barrett
National Director, James Konkel
Directors:  Stephen A. Holt, Robert E.
Emmons, Richard Weingardt, John B.
Morgan, Craig F. Robillard
*Executive Director, Sondra S. Donnel*

**1984 CECC OFFICERS:**

President, Howard C. Dutzi
Vice President, Stephen A. Holt
Secretary-Treasurer, Authur J.
Greengard, Jr.
President-Elect, Larry A. Muller
Past President, James M. Hastings
National Director, Richard C. Hepworth
Directors:  Robert V. Behrent, William
W. Moore, LaVern C. Nelson, Robert E.
Emmons, Craig F. Robillard, John B.
Morgan, Richard F. Sparlin
*Executive Director, Sondra S. Donnel*

## 1985 CECC OFFICERS:

President, Larry Muller
Vice President, Robert E. Emmons
Secretary-Treasurer, John B. Morgan
President-Elect, Stephen A. Holt
Past President, Howard C. Dutzi
National Director, Richard C. Hepworth
Directors:  Robert V. Behrent, William
W. Moore, Richard F. Sparlin, LaVern
C. Nelson, Arthur J. Greengard, Jr.,
William R. Harmon, William L. Lorah
*Executive Director, Sondra S. Donnel*

## 1986 CECC OFFICERS:

President, Stephen A. Holt
Vice President, John B. Morgan
Secretary-Treasurer, William A.
Rettberg
President-Elect, Robert E. Emmons
Past President, Larry A. Muller
National Director, Harold F. Bishop
Directors:  Arthur J. Greengard, Jr.,
William R. Harmon, William L. Lorah,
James E. Abbott, N. Kent Baker, Harold
Hollingsworth, Catherine
Kraeger-Rovey
*Executive Director, Sondra S. Donnel*

## 1987 CECC OFFICERS:

President, Robert E. Emmons
Vice President, Richard F. Sparlin
Secretary-Treasurer, William W. Moore
President-Elect, Richard Weingardt
Past President, Stephen A. Holt
National Director, Harold F. Bishop
Directors:  James E. Abbott, N. Kent
Baker, Harold Hollingsworth, Catherine
Kraeger-Rovey, Vukoslav E. Aguirre,
Kenneth J. Brotsky, Charles C. Crum
*Executive Director, Sondra S. Donnel*

## 1988 CECC OFFICERS:

President, Richard Weingardt
Vice President, William A. Rettberg
Secretary-Treasurer, Harold
Hollingsworth
President-Elect, William W. Moore
Past President, Robert E. Emmons
National Director, Howard C. Dutzi
Directors:  Vukoslav Aguirre, Kenneth J.
Brotsky, Charles C. Crum, A.S. "Andy"
Andrews, Gilbert L. Butler, Darrel V.
Holmquist, Dale D. Olhausen
*Executive Director, Sondra S. Donnel*

*Sandy Donnel, CECC Executive Director, intro-
ducing Rick "the Coach" Marshall at the April
1988 installation of officers meeting.*

*1975-76 CECC Officers and Directors, (left to right, front), Joe Meheen, Malcolm Meurer, Bill Bredar, Leonard Rice, Fu Hua Chen, Charles Meyers. (back, left to right) Russell Miller, Albert Anderson, Dave Fleming, Ronald Blatchley, Charles Bell, Kenneth Kolstad.*

*1979-80 CECC Officers and Directors (back row, left to right) Hal Bishop, Dick Hepworth, George Koonsman, Dave Austin, Lee Rice, Donavon Nickel; (front row, left to right) Wayne Irelan, Jim Hastings, Larry Muller, Bob Thompson and Ed Thorson, executive director.*

# Committee Chairmen *(partial list)*

## 1988-89

Scholarship, Ron Blatchley
Career Opportunities, Jim Harris
College Liaison, Carl Hurst
Public Relations, Gordon Meurer
CECC History, Fu Hua Chen
Membership, Lee Rice & Bob Bates
New Principals, David Stewart
Awards, Don Marek
Programs, Steve Aasheim & Bill Carley
Orientation, James McFall
CECC Statewide, Hal Bishop
Inter-Pro, Paul Pennock
Business Services, Jeff Meyers
Procurement, Bob Tallarico
Professional Liability, Jim Laraby
Construction Liaison, Mike Applegate
WBE/MBE, Richard Hepworth
Legislative, Larry Muller
Transportation, Al Menhennett
State Engineer, Chuck Reich
Environmental, Catherine Kraeger-Rovey
CEPAC, Ed Mighell
New Airport, Michael Barrett
Budget & Nominating, Bob Emmons
Long Range Planning, Craig Robillard
Professional Conduct, James Sato
CEC & Presidents Council, Bill Rettberg
Audit, John B. Morgan

## 1987-88

Legislative-State, Andy Andrews
Membership, Gilbert Butler
CEPAC, Ed Mighell
Public Relations, Jeffrey Meyers
Environmental, Richard Arber
Scholarship, Steve Aasheim
Transportation, Alan Menhennett
Procurement, Bill Rettberg
WBE/MBE, Richard Hepworth
Professional Liability, Darrel Holmquist

Program, Bob Bates
Long Range Planning, Larry Muller
State Engineer, Charles Reich
College Liaison, Carl Hurst
Career Opportunities, Cheryl Signs
Inter-Pro, William Kimball
Awards, Art Greengard
Sunset Review, Richard Hepworth
New Principals, David Stewart
Legislative-National, Mike Barrett

## 1986-87

Legislative-State, Richard Weingardt
Environmental, Geoffrey Hunkin
Awards, Stephen Aasheim
New Principals, Charles Meyers
Public Relations, Cheryl Signs
Professional Liability, Richard Hepworth
Procurement, Dale Olhausen
Inter-Pro, Robert Redwine
Program, John France
College Liaison, Phil Gerhart
CECC/CDOH Liaison, Joe Siccardi
State Engineer, Lyman Flook
Construction Liaison, Vuk Aguirre
Long Range Planning, James McFall
Membership, Stuart Monical
Legislative-National, Robert Schaevitz
Audit/Professional Conduct, Al Knott
Scholarship, Charles Crum
Education, Leslie Botham

## 1985-86

College Liaison, Charles Meyers
Awards, James Abbott
Program, Bill Rettberg
State Engineer, Lyman Flook
Scholarship, Charles Reich
CECC/CDOH Liaison, Richard Sparlin
Membership, John Williams
Education, Leslie Botham

Political Action, Harold Bishop
Professional Liability, Donavon Nickel
Professional Conduct, Robert Thompson
Public Relations, Catherine Kraeger-Rovey
Construction Liaison, William Brown
Long Range Planning, Ronald Blatchley
New Principals, Norman Almquist
Business Services, Wendell Palmer
Legislative-State, Fu Hua Chen
Urban Design, Richard Weingardt
Inter-Pro, Stanley Neujahr
Governor's Hi-Tech, Mike Barrett
Environmental, David Stewart
Procurement Practices, Robert Felsburg
Legislative-National, Al Menhennett

## 1984-85

Education, Peter Monroe
Urban Design, Richard Weingardt
Inter-Pro, Robert Clay
Membership, Don Marek
CECC/CDOH Liaison, Richard Sparlin
State Engineer, Leonard Rice
Public Relations, Bill Rettberg
Construction Liaision, Vuk Aguirre
Legislative-State, Perry Tyree
PAC, Harold Bishop
Scholarship, Harlan Erker
ACEC Conference, Kenneth Kolstad
Program, Gordon Meurer
College Liaison, Charles Meyers
Audit, William Hawes
Long Range Planning, Ron Blatchley
Awards, Alan Huggins
Professional Conduct, Jack Haney

## 1983-84

Business Services, Eugene Zimmer
Construction Liaison, Vuk Aguirre
Education, James McFall
Inter-Pro, Howard Browning
Environmental, James Abbott
Legislative-State, Leonard Rice
PAC, Oliver Watts
Legislative-National, John Haley
Public Relations, Harold Bishop
Long Range Planning, George Koonsman
Membership, Joe Meheen
State Engineer, Richard Hepworth
College Liaison, Ronald Hensen
Professional Conduct, Donald Preszler
Program Committee, Art Greengard

Public Relations, Bill Rettberg
Awards, Stanley Neujahr
Scholarship, Curtis McMillin
Procurement Practices, Elmer Claycomb
Colorado Design Alliance, John Park

## 1982-83

Budget, Richard Hepworth
Audit, George Koonsman
Business Services, Eugene Zimmer
Construction Liaison, Alan Huggins
Inter-Pro, Larry Gambrell
Education, James McFall
Environmental, James Abbott
Legislative-State, Wayne Irelan
Legislative-National, Cheryl Signs
CEPAC, Ronald Blatchley
Long Range Planning, George Cogorno
Membership, Harold Bishop
College Liaison, Theodore Zorich
Professional Conduct, Donald Prezler
Program, Joe Meheen
Public Relations, Bob Emmons
Awards, Art Greengard
Scholarship, Robert Kroening
Procurement Practices, Bernard Poppenga
Colorado Design Alliance, James Borman

## 1981-82

Audit, James Konkel
Education, John Haley
Inter-Pro, Steve Schermerhorn
Legislative-State, Donald Ricketts
Environmental, James Abbott
Legislative-National, Cheryl Signs
Long Range Planning, Harold Bishop
Membership, Frank Holliday
College Liaison, Tom Pitts
Professional Conduct, William Bredar
Program, Robert Leigh
Public Relations, Bob Emmons
Awards, Ken Baker
Scholarship Committee, Jack McKee
Procurement, William Blackmer
Computer Graphics, Richard Sparlin

## 1980-81

Audit, Dick Fleming & Frank Holliday
Construction Liaison, Gilbert Rindahl
Business Services, Gordon Meurer

Environmental, Robert Leigh
Legislative-State, Stephen Holt
Legislative-National, Lyman Flook
Long Range Planning, Mike Barrett
Membership, Allen Wassenaar
College Liaison, Larry Faulkner
Professional Conduct, William Bredar
Program, Robert McWhinnie
Public Relations, Richard Weingardt
Scholarship, Don Marek
Awards, D.E. Burroughs
Procurement Practices, Thomas Norton

## 1979-1980

New Members, Gordon Meurer
Awards, Al Menhennett
College Liaison, Larry Faulkner
Scholarship, Don Bressler
Professional Conduct, W.L. Bredar
Procurement Practices, Mike Barrett
Long Range Planning, Donald Preszler
Program, John Haley

## 1978-79

Business Services, Kent Baker
Education, Robert Voiland
Construction Liaison, Jack Haney
Environmental, William Bredar
Legislative-State, Leonard Rice
Legislative-National, Victor Weidmann
CEPAC, Ronald Blatchley
ACE/PAC, Joe Meheen
Inter-Pro, G.V. Kohnert
Program, James Konkel
Professional Conduct, Al Anderson
Public Relations, Larry Muller
Long Range Planning, Al Anderson
College Liaison, Charles Meyers
Nominating, Ronald Blatchley
Scholarship, Theodore Zorich
Continuing Education, George Cogorno
Procurement Practices, Al Menhennett

## 1977-78

Membership, Larry Muller
Legislative-State, Leonard Rice
Business Services, Kenneth Kolstad
Education, David Austin
Environmental, Robert McWhinnie

Legislative-National, Victor Weidmann
Competition Reaction, Russell Miller
Long Range Planning, Charles Meyers
Professional Conduct, Richard Kellogg
Program, John Bright
Awards, Donald Preszler
Scholarship, Thomas Young

## 1976-77

Membership, Richard Hepworth
Competition Reaction, Morton Bittinger
Public Relations, Frank Holliday
Long Range Planning, James McFall
Inter-Pro, Jack McKee
Professional Conduct, Ib Falk Jorgensen
Legislative-State, Edward Thorson
Education, Don Marek
Scholarship, Sol Flax
Legislative-National, Victor Weidmann
Environmental, Harold Bishop
Construction Liaison, Dean Peterson
Business Services, Howard Dutzi
Awards, Mike Barrett
Program, Gordon McDonald

## 1971-72

Audit, Robert Voiland
Budget & Nominating, Ralph Becker
Long Range Planning, Kenneth Wright
Membership, Jack Lovell
Professional Conduct, Olin Kalmbach
Programs, Andrew Pfeiffenberger
Public Relations, Ib Jorgensen & Don Marek
Scholarship, James Stewart
Awards, Carl Ray, Jr.
Business Services, Kenneth Kolstad
Education & Recruiting, Sol Flax
AE Relations, Albert Anderson
Environmental Design, Allen Menhenett
Legislative-National, David E. Austin
Legislative-State, Willard Quirk
Manuals of Practice, Sebastian Ruma
Olympics '76, Robert E. Barber
Construction Cooperative, David Fleming
Liaison with AGC, Orley O. Phillips

## 1970-71

Education, James Stewart
Budget, Kenneth Wright
Membership, Robert Behrent

AE Relations, Vern Winkel
Public Relations, William Bredar
Olympics '76, David Fleming
Legislative-National, W.A. Clevenger
Long Range Planning, Kenneth Kolstad
Legislative-State, James Konkel
H.B. 1100, S.T. Thorfinnson
Nominating, Kenneth Wright
Professional Conduct, Kenneth Wright
Past Presidents, William Hawes
Private Practice, Orley Phillips

## 1969-70

Ethical Practices, Fu Hua Chen
Environmental, Roy Russell
Manual Practices, C.H. Honaker
Public Relations, S.T. Thorfinnson
Membership, Vernon Winkel
Legislative-State, Donald Preszler
Private Practice, James Konkel
Long Range Planning, W.A. Clevenger
Legislative-National, George Koonsman
Education, Don Marek
Awards, William Bredar
Auditing, Rhuel Andersen
Budget, William Hawes
Constitution and By-Laws, Ken Kolstad
Government Competion, Carl Ray
Nominating, William Hawes
Program, Mike Barrett

## 1968-69

Private Practice, A.D. Swanson
Legislative-State, Fred Rink
Ethical Practices, W.S. Langebartel
Constitution and By-Laws, Charles Fisk
Program, Charles Meurer
Public Relations, Mike Barrett
AE Relations, R.W. Becker
Manual Practices, Dwight Sayles
Symposium, S.T. Thorfinnson
Legislative-National, Robert Lord, Jr.
Membership, Fu Hua Chen
Education, Donavon Nickel

## 1967-68

Constitution & By-Laws, Earl Heckman
Defense Panel, Orley Phillips
AE Relations, R.W. Becker
Inter-Pro, Kenneth Wright
Budget, W.A. Clevenger
Legislative-National, Donald Preszler
Long Range Planning, William Hinch
Membership, William Hawes
Private Practices, Alvin Swanson
Public Relations, Mike Barrett
Awards, Mike Barrett
Education, George Koonsman
Manual of Practice, Dwight Sayles
Professions, E. Vernon Konkel
Program, Kenneth Wright
Legislative-State, John Bunts

## 1965-66

Inter-Pro, Kenneth Wright
Public Relations, Kenneth Wright
Long Range Planning, W. Langebartel
Membership, Robert Rice
Programs, Robert Behrent
Legislative-State, Donald Preszler

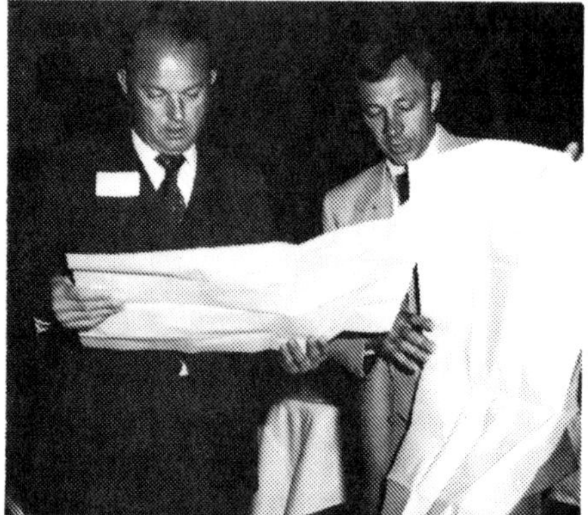

*Dale Olhausen (right) & Wayne Irelan
at April 1981 CECC meeting.*

# Appendix C

# Engineering Excellence Awards

The Annual Engineering Excellence Awards Program was initiated in 1967 to recognize those engineering achievements demonstrating the highest degree of merit and ingenuity, and providing a major contribution to technical, economical or social advancements.

## 1988 AWARDEES

Camp Dresser & McKee International
Honduras:  Solving a Critical Water
Shortage

Richard Weingardt Consultants, Inc.
City Park Recreation Complex,
Westminster

Harza Engineering Company
Blue River Hydro Project, Dillon

RG Consulting Engineers, Inc.
Clarke Farms Wellhouse and Treatment
Facility, Parker

Black & Veatch Engineers-Architects
Gunnison, Colorado Wastewater Treatment Improvements

David J. Love & Associates, Inc.
Boulder Creek Corridor - 5 Mile Path &
Waterway

J.F. Sato and Associates, Inc.
Heart Butte Dam Modification Project,
Bismarck, North Dakota

## 1987 AWARDEES

Bishop, Brogden & Rumph, Inc.*
Niobrara River Water Development Plan,
Nebraska

Richard Weingardt Consultants, Inc.
United Airlines Concourse B, Stapleton
International Airport

HDR Infrastructure, Inc.
Sunshine Hydroelectric Facility, Boulder

Brown and Caldwell
Sludge Management System at Colorado
Springs

URS Corporation
Garden of the Gods Road/Railroad
Separation

Daniel, Mann, Johnson & Mendenhall
Barker Dam Stabilization, Nederland

## 1986 AWARDEES

Stewart Environmental Consultants, Inc.
Larimer County Non-Hazardous Liquid
Waste Treatment Study

Richard Weingardt Consultants, Inc.
Renovation of Mount View Acres, Aurora

J.F. Sato and Associates, Inc.
Marine Thruster as a Hydroelectric Turbine
Eagle Creek National Fish Hatchery, Estacada, Oregon

Black & Veatch, Engineers-Architects
Metro Sewage District Congeneration Facility, Denver

KKBNA, Inc., Consulting Engineers
Snake River Water Treatment Facility, Keystone

McCarty Engineering Consultants, Inc.
Bennett Street Improvement, Bennett

## 1985 AWARDEES

Black & Veatch, Engineers-Architects
Sludge Management Plan for Metropolitan Denver

Morrison-Knudsen Engineers, Inc.*
Middle Fork Dam, Parachute

CH2M Hill, Inc.
Denver Potable Water Re-Use Demonstration Plant

Centennial Engineering, Inc.
Stapleton International Airport Traffic Signal System

URS Company, Denver Regional Office
Georgetown Loop Bridge Reconstruction

## 1984 AWARDEES

Wright Water Engineers, Inc.
Hydrological Study-Angel Fire Litigation, Raton, New Mexico

Chen and Associates
Rio Grande Dam Rehabilitation, San Luis Valley Irrigation District

Richard Weingardt Consultants, Inc.
92nd Avenue Overpass and Roadway, Westminster

LONCO, Inc.
United Airlines Underground Baggage Facility, Stapleton

## 1983 AWARDEES

Centennial Engineering, Inc.
Application of Slurry Filters to Cement Manufacturing, Amarillo, Texas

Camp Dresser & McKee, Inc.*
Bishop's Lodge Wastewater Reclamation Facility, Santa Fe, New Mexico

Arix Engineers, Architects*
Wastewater Treatment Plant, Sterling

M & I, Inc.
Vail Wastewater Treatment Facility

CH2M Hill
Foothills Water Treatment Plant, Denver

## 1982 AWARDEES

Coury and Associates, Inc.
Geothermal District Heating System for the Town of Pagosa Springs

URS Company*
Salinity Investigation of the Glenwood-Dotsero Springs Control Unit

McFall, Konkel & Kimball Consulting Engineers
United States Post Office, Aspen

ARIX, A Professional Corporation
Jackson Wastewater Treatment Plant

Harza Engineering, Inc.
Strontia Springs Diversion Dam

**1981 AWARDEES**

Wright-McLaughlin Engineers
The Northglenn Water Management
Program

McFall, Konkel & Kimball Consulting
Engineers, Inc.
Obermeyer Corporate Offices

CH2M Hill*
Ray D. Nixon Power Plant Water Systems

**1980 AWARDEES**

CH2M Hill
Water Reuse Demo Plant, Denver

Richard Weingardt Consultants, Inc.
Skylights for Helen Bonfils Theatre,
Denver

Richard Weingardt Consultants, Inc.
American National Bank Parking Structure, Denver

Richard Weingardt Consultants, Inc.
International Athletic Club, Denver

KKBNA, Inc., Consulting Engineers
Denver Center for Performing Arts

**1979 AWARDEES**

Centennial Engineering
Coal Firing System for Cement Kilns,
Clarkdale, Arizona

DMJM-Phillips, Reister, Haley/Gruen
Associates
Design Development I-70 Through
Glenwood Canyon

Richard Weingardt Consultants, Inc.
Mount Olivet Administration Building,
Denver

Centennial Engineering
Relocation of a Section of Colorado Highway No. 91

LONCO, Inc.
Bridge Building, Pedestrian Corridors at
Stapleton

**1978 AWARDEES**

DMJM-Phillips, Reister, Inc.*
Movable Stands, Denver Mile High Stadium

CH2M Hill*
Bioconversion Plant, Lamar

KKBNA, Inc., Consulting Engineers
Portland Oregon Tri-Met Light Rail Transit System

Meheen Corporation
Structure for Runway, Stapleton

**1977 AWARDEES**

Walton-Abeyta and Associates, Inc.
Solar Energy Report for Burdines, Miami,
Florida

KKBNA, Inc., Consulting Engineers
Evolution of a Community, Eagle River
Valley of Colorado

Black & Veatch
Green Mountain Pump Station and
Reservoir

KKBNA, Inc., Consulting Engineers
The McNichols Sports Arena, Denver

Rice, Marek, Holtz & Patterson, Inc.
Solid Waste and Pathological Incinerator,
Hamilton, Montana

**1976 AWARDEES**

Chen and Associates
Rehabilitation of a School in Arapahoe
County

KKBNA, Inc., Consulting Engineers
Stadium Roof for the University of Idaho

Wright-McLaughlin Engineers
Development on the South Platte River,
Denver

Leonard Rice Consulting Water Engineers
Urban Drainage at Colorado State University
sity

Leonard Rice Consulting Water Engineers
Northeast Heights Drainage Management
Plan, Albuquerque, New Mexico

**1975 AWARDEES**

Wright-McLaughlin/Jorgensen-Hendrickson
drickson
Water Storage and Distribution System,
City of Arvada

International Engineering Company, Inc.
Research of Precast Concrete Retaining
Wall over Vail Pass

KKBNA, Inc., Consulting Engineers
Hennepin County Building, Minneapolis

R. Keith Hook and Associates, Inc.
Water Quality Control System, Colorado
Springs

McFall and Konkel, Consulting Engineers
Heating and Ventilating System, Denver's
Park Central Project

**1974 AWARDEES**

Wright McLaughlin Engineers, Inc.*
Long Range Wastewater Management
Plan for Cleveland, Ohio

KKBNA, Inc., Consulting Engineers
Mile Hi Church of Religious Science,
Lakewood

Technical Service Company
Detroit News Sterling Heights Plant,
Detroit

Leonard Rice Consulting Water
Engineers, Inc.*
Boulder Creek Gravel Extraction

David E. Fleming Company and Parsons,
Brinckerhoff, Quade & Douglas, Inc.
Computer Simulation Model (Corsim II),
Colorado and White Rivers

**1973 AWARDEES**

KKBNA, Inc., Consulting Engineers*
TWA Hanger Structure, Kansas City

Nelson, Haley, Patterson and Quirk
Urban Transit Design Study, Denver

Black & Veatch
Cheesman Valve Structure, Deckers

URS/The Ken R. White Company
Christo Valley Curtain, Rifle

**1972 AWARDEES**

Wright-McLaughlin Engineers
Denver Botanic Gardens Water Facility

R.V. Lord and Associates, Inc.
Lake of Pines Sewage Disposal System

Behrent Engineering Company
NCAR Particle Control Chamber

Henningson, Durham & Richardson, Inc.
Grand Junction Water Pollution Control

KKBNA, Inc., Consulting Engineers
Denver Center Office Building

## 1970 AWARDEES

Barber-Nichols Engineering Company
Lear Engine

Black & Veatch
Shanahan Reservoir

Henningson, Durham & Richardson
Westminster Water Plant

KKBNA, Inc., Consulting Engineers
Space Frame Roof System

The Ken R. White Company
Twin Arch Bridges in Washington

## 1969 AWARDEES

Wright-McLaughlin Engineers
Urban Storm Drainage Criteria Manual

McFall-Konkel & Kimball Consulting
Engineers
Financial Programs Building

Rice-Marek-Harral & Associates
King Soopers Blast Freezing Facility

Henningson-Durham & Richardson
Grand Junction Water Treatment Plant

## 1968 AWARDEE

KKBNA, Inc., Consulting Engineers
Denver Currigan Convention Center

## 1967 AWARDEE

Ken R. White Company
Ideal Cement Company Plant, Seattle

*Indicates projects also receiving ACEC national Awards.*

---

## 1989 AWARD WINNERS

*The following winners will be announced at the 23rd Annual Awards Banquet during National Engineers Week, February 1989*

*David L. Adams Associates,
Union Colony Civic Center, Greeley*

*Centennial Engineering,
Littleton Railroad Depression*

*CH2M Hill, Monument Creek Regional Wastewater Treatment Plant*

*McLaughlin Water Engineers, Inc.,
Little Dry Creek Flood Control Improvements, Englewood*

*J.F. Sato and Associates, Inc.,
Application of Global PositioningSystem Satellite Technology, Colorado Springs*

*Richard Weingardt Consultants, Inc.,
Gerald Ford Amphitheater, Vail*

# Honor Awards

**ORLEY PHILLIPS AWARD**

1980, Orley Phillips

1982, Fu Hua Chen

1983, Malcolm Meurer

1984, James Konkel

1987, Leonard Rice

**LEGISLATOR OF THE YEAR**

1986, Representative Donald Mielke

1987, Representative Paul Schauer

1988, Senator David Wattenberg

**GENERAL PALMER AWARD**

1988, George Wallace

1989, Jack Perlmutter

**NEW PRINCIPAL OF YEAR**

1986, 1987, David Stewart

1988, Gregg Ten Eyck

**MEDIA PERSON OF THE YEAR**

1988, Rosalie Merzbach

# Sports Awards

**Golf 1979**

Low gross, Franklin Blake
Low net, James Hastings

**Golf 1980**

Low gross, Franklin Blake & Albert Anderson
Low net, James Hasting

**Golf 1981**

Low gross, Al Anderson
Low net, Jack Blake

**Golf 1982**

Low gross, Al Anderson
Low net, Sebastian Ruma

**Golf 1983**

Low gross, Al Anderson
Low net, Gordon Meurer

**Golf 1984**

Low gross, Jack Blake
Low net, Sebastian Ruma

**Golf 1985**

Low gross, Jim Hastings
Low net, Gordon Meurer

**Golf 1986**

Low gross, Donavon Nickel
Low net, James Konkel

**Golf 1987**

Low gross, Lawrence Boval
Low net, James Hastings

**Golf 1988**

Low gross, William Artist
Low net, Vern Nelson & Dale Olhausen

**Tennis 1983**

Men's Doubles, Rich Weingardt & John Williams
Mixed Doubles, John & Carole Morgan
Men's Singles, George Koonsman

**Tennis 1984**

Men's Doubles, John Morgan & George Koonsman
Mixed Doubles, John & Carole Morgan

**Tennis 1985**

Men's Doubles, John Morgan & George Koonsman
Mixed Doubles, John & Carole Morgan

**Tennis 1986**

Men's Doubles, Rich Weingardt & John
Williams
Mixed Doubles, Rich Weingardt &
Sandy Moore

**Tennis 1987**

Mixed Doubles, Eugene & Sharon
Zimmer

**Tennis 1988**

Mixed Doubles, Dave & Mary Stewart

*Golf winners (l to r), Gil Butler, Vern Nelson,
Dale Olhausen, July, 1988.*

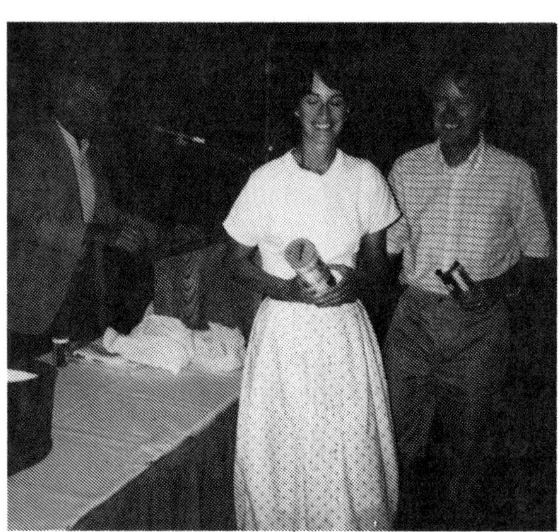

*New "tennis stars," Julie & Bill Rettberg. Lee Rice
tennis chairman is at left.*

*Harold Hollingsworth was golfer closest to the
pin, July, 1988.*

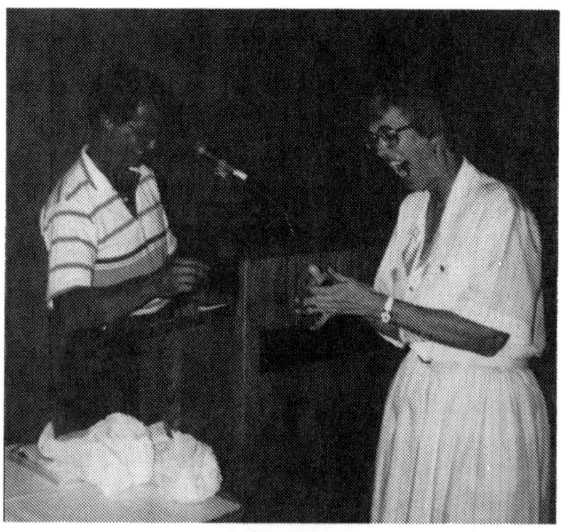

*Bill Artist, golf tournament chairman, presents
Nancy Butler with her "prize".*

# Locations of CECC Conferences

1988, Sheraton Hotel, Steamboat Springs

1987, Hilton Hotel, Breckenridge

1986, Aspen Lodge in Estes Park

1985, The Snowmass Club in Snowmass

1984, Garden of the Gods Club,
Colorado Springs

1983, Keystone Resort, Keystone

1982, Garden of the Gods Club,
Colorado Springs

1981, Sheraton Hotel, Steamboat Springs

1980, The Mark Hotel, Vail

1979, Perry Park Country Club,
Perry Park

1978, Antlers Hotel, Colorado Springs

*Jim McFall has attended all but one CECC state convention.*

*Nothing stops Al and Bobbie Knott from attending conference.*

*CECC's 1988 annual state convention was held at the Sheraton conference center, Steamboat Springs, July, 28-31.*

# **Meeting Programs** (partial list)

The history of any association and the concerns it was experiencing are reflected in its programs (and speakers).  The following are those CECC had over the years:

1988-89

May, Bill Smith, P.E., "Progress of New Denver, Colorado Airport"
June, Jim Bryant, "Colorado & the Superconductor Super Collider"
July, Dag Knudsen, "Winning Presentations," and Denny Kercher,
"Managing Stress and Burnout"
August, A. Ray Chamberlain, P.E., "State Transportation Mobility Plan"
September, James Poirot, P.E., "the Future of Engineering & ACEC"
October, Lt. Governor Michael Callahan, "Always Buy Colorado"
November, Daniel Smith, "Recent Failures in Engineering & Liability Issues"
January, August Perez & Company, "Architectural Design of New Airport"

1987-88

July, Dr. Edward Cooper, "Marketing in a Recessionary Economy"
August, John Mrozek, Bill Smith, "Denver's New Airport-A Status Report"
August, Lowell Jackson, "Colorado's Highway Projects & Financing"
October, Dr. Richard Hoerl, "Future Trends and Long Range Planning"
November, Representative Paul Schauer, "Legislative Update"
January, Susan Quinn, "Quality Professional Services"
February, David Baskett, "An Update on RTD's 1988 Plans"
March, Leon Wurl, "W-470 Status Update"

1986-87

June, Representative Don Mielke, "Legislative Update"
September, Lester Smith, "Contract Indemnification Clauses & Peer Review"
October, Ray Bullock, "Funding of Civil Works Projects in an Era of Reduced
Government Spending"
November, Richard Seebass,  Francis Kulacki, and John Golden, "Issues in Colorado
Engineering Education"
December, Robert Brown, "Colorado 100 Years Ago"
January, David Baskett, "Implementing our Regional Transit System"
March, Uli Kappus, "Strategy for Financing Colorado's Water Future"

1985-86

Les Poggemeyer, "ACEC and the Future"
Walt Borneman, "Georgetown P.R. Loop Reconstruction"
Richard Lopez, "Proposed New Denver Airport"
Bill Miller, "System-Wide Environmental Impact Statement"
Bill Smith, "Expansion of Existing Airport"
John Mrozek, "Public Works Projects of the Future"
Bill Lamont, "Future Planning of Denver into the 2000's"
Uli Kappus, "The Role of Water & Power Authority"


1984-85

Karen Sussman, "Stress"
Stephen Andrade, "High Tech in Colorado"
Roger Wiggins, "Personal Time"
Fu Hua Chen, "Engineering in China"
Mary DeLapp, "Grapho-Analysis"
Roy Romer, "The Political Future"
Bill Bredar, "Professional Liability"
Pete Smythe, "Anecdotes"


1983-84

August, Evelyn Hottenstein, "Making Professional Presentations with Pizzazz"
September, Kathy Frazer, "The Secrets of Marketing Engineering Services"
October, Jean Ackerman, "Colorado Business Trends of the 80's"
November, Jerry Quinn, "Does the Left Hand Know What the Right is Doing?"
December, Harriet Hagestad, "Visual Image for Engineers and Spouses"
January, Ken Johnson, "Computers, Friend or Foe?"
March, Representative Don Mielke, "Legislative Involvement of Engineers"
May, Dr. Richard Brandenburg, "Executive M.B.A.'s"


1982-83

"Presentation by National Industrial Committee," Panel
Judy Riggenbach, "Body Language"
"To Bid or Not To Bid," Panel
Everett Thompson, "Managing of Human Resources"
"Ownership Transfer, To Sell, Buy or Go Public," Panel
"Marketing Professional Services," Panel
"Self Insurance," Panel

1981-82

"Coming Electrical Energy Shortage in Colorado"
"Public Education Regarding Profits in Business"
"Public Relations"
"The Colorado Front Range Project"
"Inter-professional Relations with Lawyers and Architects"
"Proposed Consolidation of ACEC and PEC"
"Arbitration Disputes in Engineering and Construction"

1980-81

"World Energy Outlook"
"Estate Planning"
"Employee Motivation"
"Business Writing"
"Project Management and Reporting"

1979-80

June, R. Rautenstraus, "Engineering Technology of the Future"
September, E. Howell, "Who is Winning the Professional Liability Game"
October, L. Taylor, "The Energy Problem's Impact on Colorado's Future"
November, R. Specht, "Outlooks for Engineering and Construction in the Region"
January, L. Baldwin, "Bottlenecks in the Supply Lines"
March, G. Barnes, "Activities on Capitol Hill and Across the Nation"
March, J. Doussard, "Estate Planning, Pension and Profit Sharing"

1978-79

"Trip to Nuclear Power Plant"
"Discussions on Energy"
Michael Hough, "Action Plan for Success"
"CEC/Colorado Credit Union Possibilities"
"Affirmative Action Programs"
Fu Hua Chen, "China Trip"
Report by Duane Monical, ACEC President

1977-78

"Business Ethics"
"The Law Relating to Joint Ventures"
"Motivation and Goal Setting"
"Tax Laws"
"Liability and Loss Prevention"
"Re-Certification of Engineers"

1976-78

"Energy for the Year 2000"
"Proposed New Engineering Registration Law"
"Consultants Liability Problems"
"Loss Prevention Seminar"

1971-72

"Professional Employment"
"The Effect of Government Influences"
"Building Codes"
"The Engineer-Client Relationship"
"Professional Liability"
"The Future Business Picture"
"Organizational Structure of CECC"

1968-69

May, Report on CEC/US May Convention
June, Representatives from A.G.C.
August, "Senator Peter Dominick, Guest Speaker"
September,  Wright and J. Konkel, "Environmental Discussion"
October, Kuhler, "Mass Transit"
January, "Discussion on Proposed Constitutional Changes"
March, "Discussion of the Code of Professional Conduct"

*Incoming president, Dick Hepworth is congratu-
lated by Mike Barrett (left) as Hal Bishop (seated)
applauds, 1981.*

*Governor Roy Romer (left) reviewing CECC's
booklet of engineers willing to serve on boards and
commissions with Steve Holt, 1986.*

1967-68

May, Glenn Saunders, "Finding the Right Expert Witness is Like Finding the Right Wife"
June, Ken Burrus, "How to Increase Profit with Better Accounting Procedures"
August, Manufacturer's representatives, "Planning, Specifying, and Using Package Units"
October, Glenn Saunders & Jack Ross, "The Consulting Engineer in Court"
November, "State Technical Services"
December, "Combatting Obsolescence in Engineers - Improving your Reading Speed"
January, "Proposed Code of Ethics"

1965-66

May, Charles Gover, "Union Problems"
June, Panel, "Environmental Design"
August, John Felker, "Review of Group Insurance"
September, Col. Henry Ochs, Jr., "New Engineering Registration Law for Colorado"
October, Panel, "Latest Developments of Unions As They Affect CECC"
November, Mr. Howell, "Liability Insurance"
January, Dr. Irvin Hendryson, "Government and the Profession"
March, R.J. DeLaCastro & Pierre DuBois, "Common Problems Regarding the General Contracting Association and CEC

*ACEC President Russ Smith with Mike Barrett (right), 1983.*

*CECC scholarship winner, Jim Brugger, and Jim Hastings (right), 1984.*

*Distinguished panel of judges: (left to right); (bottom) Curt Dale, president AIA Denver; Dwight Bower, CDOH Deputy Director; A. Richard Seebass, Dean of College of Engineering, University of Colorado; Bill Smith, Assistant Director, New Denver Airport Authority, (top) Jeff Rundles, Editor Denver Business Journal; State Representative Paul Schauer; Cal Cox, president of AGC Colorado. Not shown, Ernie Bjorkman, News Anchor channel 7 T.V. (photography by Ed Bernstein).*

*Bill Smith rates project merits.*

*Dwight Bower checks details of project.*

*Jim McFall (foreground left) and Malcolm Meurer (right) were overheard discussing how things had changed since they have been in the consulting business. McFall commented, "Our client nowadays, whether private or government, is a committee rather than a single headman. The take-charge-I'll-get-it-done type of individual has vanished." Meurer agreed and added, "The days when a contract could be signed on the back-of-an-envelope are also gone."*

# Appendix D

# ACEC Member Relationship

American Consulting Engineers Council is a federation of associations serving the needs of consulting engineers on a metropolitan area, state or regional basis. These Member Organizations maintain a great degree of independence, electing their own officers, electing or appointing the Directors and Alternate Directors that constitute the ACEC governing body, and conducting service programs that are responsive to the needs of their members. A strong spirit of "States' Rights" has always characterized CEC/US and ACEC.

National affiliation criteria require that a Member Organization "have purposes consistent with those of the (national) Council, and...a constitution, bylaws and activities which do not conflict with those of the Council..." These requirements were purposely left somewhat flexible in order to accommodate a variety of minor differences in the organizational documents, policies and procedures of the highly independent groups joined to form ACEC. Each Member Organization collects dues from its Member Firms, including, national dues remitted to ACEC headquarters, and an amount for the support of local activities.

# ACEC Offices Held by CECC Members

Orley O. Phillips, Vice President 1959-60

Eugene B. Waggoner, Vice President 1962-63, President-Elect 1965-66, President 1966-67

E. Vernon Konkel, Vice President 1964-66

Malcolm R. Meurer, Secretary-Treasurer 1969-70, President-Elect 1973-74, President 1974-75

William A. Clevenger, Vice President 1971-73

Fu Hua Chen, Sr. Vice President 1981-83

Leonard Rice, Vice President 1985-87

James W. Poirot, President-Elect 1988-89, President 1989-90

## MEURER, ACEC PRESIDENT 1974-1975

MALCOLM R. MEURER was raised and educated in Indiana. He was born June 2, 1918 in Sullivan, Indiana. He graduated from Rose Polytechnic Institute in Terre Haute, Indiana, in 1949 with a Bachelor of Science degree in Civil Engineering. During World War II, from 1943 to 1946, he served in a heavy artillery unit of the U.S. Army. After being discharged he returned to school and following graduation became a member of a construction company in Grand Rapids, Michigan, which was engaged in heavy construction. In 1953 he was employed by the firm of Drury, McNamee and Porter, consulting engineers, in Ann Arbor, Michigan.

Meurer is one of three original principals in the firm of Meurer, Serafini and Meurer, Inc., organized in 1955 in Denver. Of the firm's many civil works projects constructed in Colorado, he is especially proud of their bridge over the Colorado River at Palisade. In 1963 they designed the Arvada Water Treatment Plant and the first Micro-flac horizontal filter beds used in Colorado.

In 1974, he was appointed to President Ford's Economic Council. He received the Colorado Engineering Council's Award of Honor in 1980 and the Orley Phillips award in 1983. A fellow of ACEC he is both a registered professional engineer and a land surveyor in Colorado. He is also a member of the National Society of Professional Engineers and the American Society of Civil Engineers.

Malcolm resides in Denver with his wife, Queen. The Meurers have three children.

## POIROT, ACEC PRESIDENT 1989-1990

JAMES W. POIROT was born in Douglas, Wyoming, November 24, 1931. He moved to Colorado in 1983 when he became the Chairman of the Board for CH2M Hill.

At Oregon State University he earned a B.S. degree in Civil Engineering, and graduated in 1953. Poirot was a design/project engineer in CH2M Hill's Corvallis, Oregon office where he first joined the firm. Before becoming Chairman of the Board, he served three years as District Manager for the Eastern District. Prior to this, he headed the Northwest District for six years, two of which he simultaneously served as District Manager of the Southwest District. He served as an officer in the U.S. Army Corps of Engineers, assigned to the construction of Eielson Air Force Based near Fairbanks, Alaska.

Poirot received ENR's Construction "Man of the Year" award in February, 1988 to recognize his three-year effort as Chairman of ASCE's Quality Manual Steering Committee, leading the development of a landmark manual on building quality into construction. He also received "Those Who Made Marks" in 1985 and 1987, presented by ENR to honor individuals who serve the best interests of the construction industry.

In 1985 he received the ASCE "Edmund Friedman Professional Recognition" award, and ASCE-Colorado's "Civil Engineer of the Year" award.

Jim is married to Raeda and has two grown sons, Ron and Steve.

## FELLOWS OF ACEC

**Fu Hua Chen**
**David E. Fleming**
**James H. Konkel**
**George L. Koonsman**
**Malcolm R. Meurer**
**Orley O. Phillips**
**Victor H. Weidmann**

*CECC members at Boston ACEC Conference, 1985, (left to right) Art Greengard, Fu Hua Chen, Howard Dutzi, Jim Hastings, Don Berry, Sandy Donnel, Karen Anne Jones (University of Colorado student) and Hal Bishop.*

# Current CECC Roster

**MEMBER FIRMS**

A-E DESIGN ASSOCIATES, P.C., Fort Collins

A.C. ENGINEERING COMPANY, Denver

AD & C GROUP, INC., Lakewood

DAVID L. ADAMS ASSOCIATES, INC., Denver

ADAMS-AVERY, LTD., Denver

ADG ENGINEERING, INC., Englewood

AE CONSULTANTS, INC., Denver

AGUIRRE ENGINEERS, INC., Englewood

ANDERSON AND HASTINGS, Denver

APPLIED ENVIRONMENTAL, INC., Englewood

RICHARD P. ARBER ASSOCIATES, INC., Denver

ARCHITECTURAL ENGINEERING ASSOCIATES, Greeley

ARIX CORPORATION, Greeley

ARMSTRONG CONSULTANTS, INC., Grand Junction

ARNESEN & ASSOCIATES, INC., Broomfield

ATC ENGINEERING CONSULTANTS, INC., Englewood

ATKINSON-NOLAND & ASSOCIATES, INC., Boulder

THE BAKER REGAN GROUP, INC., Greeley

BALLOFFET AND ASSOCIATES, INC., Fort Collins

BANNER ASSOCIATES, INC., Grand Junction

R.W. BECK AND ASSOCIATES, Denver

BEHRENT ENGINEERING COMPANY, Denver

CHARLES E. BELL/CONSULTING ENGINEERS, Englewood

BERGE-BREWER & ASSOCIATES, Colorado Springs

BISHOP-BROGDEN ASSOCIATES, INC., Lakewood

BLACK & VEATCH, Aurora

BLATCHLEY ASSOCIATES, INC., Denver

BORMAN/SMITH/BUSH & PARTNERS, INC., Denver

CHARLES C. BOWMAN ASSOCIATES, INC., Boulder

BOYD BROWN STUDE & CAMBERN, Englewood

BOYLE ENGINEERING CORPORATION, Lakewood

BROWN AND CALDWELL, Denver

BUCHER, WILLIS & RATLIFF, ColoradoSprings

BUCKHORN GEOTECH, INC., Montrose

BURKE ASSOCIATES, INC., Grand Junction

C4 CONSTRUCTION SOFTWARE CONSULTANTS, Denver

CAER ENGINEERS, INC., Golden

CAMP DRESSER & MCKEE, INC., Denver

CARROLL & LANGE, INC., Lakewood

GERALD CASPE, CONSULTING ENGINEER, Denver

CATOR, RUMA & ASSOCIATES CO., Lakewood

CENTENNIAL ENGINEERING, INC., Arvada

CH2M HILL, Denver

CHEN-NORTHERN, INC., Denver

CHERYL SIGNS ENGINEERING, Denver

CIVIL DESIGN CONSULTANTS, Steamboat Springs

CLAY & ASSOCIATES, LTD., Denver

GLENN A. CLAYDEN & ASSOCIATES, Littleton

J.T. COLLINS AND ASSOCIATES, INC., Boulder

CONSULTING ENGINERS, INC., Colorado Springs

CONTRA, LTD., Arvada

COURY AND ASSOCIATES, INC., Wheat Ridge

GARLAND D. COX ASSOCIATES, INC., Denver

CREATIVE STRUCTURES, INC., Denver

CRS SIRRINE INFRASTRUCTURE GROUP, Denver

CTL-THOMPSON, INC., Denver

CUSTOM ENGINEERING, INC., Englewood

DAMES & MOORE, Golden

DANZBERGER AND ASSOCIATES, Golden

DEL-MONT CONSULTANTS, INC., Montrose

DENVER DESIGN ASSOCIATES, Denver

DENVER ENGINEERING ASSOCIATES, INC., Denver

DESIGN STRUCTURES, INC., Denver

DICKINSON-ASSOCIATES, INC., Evergreen

DMJM, Denver

DONOHUE & ASSOCIATES, INC., Fort Collins

HOWARD C. DUTZI & ASSOCIATES, INC., Colorado Springs

ELECTRICAL POWER CONSULTANTS, INC., Lakewood

ELECTRICAL SYSTEMS CONSULTANTS, INC., Fort Collins

EMK CONSULTANTS, INC., Englewood

EMPIRE LABORATORIES, INC., Fort Collins

ENGINEERING PROFESSIONALS, INC., Fort Collins

ENGINEERING SERVICE COMPANY, Aurora

EPPS & ASSOCIATES CONSULTING ENGINEERS, Englewood

ERNST ENGINEERING COMPANY, Durango

EVANS, KUHN & ASSOCIATES, INC., Englewood

FACILITECH, INC., Lakewood

FELSBURG, HOLT AND ULLEVIG, Englewood

FIGG AND MULLER ENGINEERS, INC., Denver

SOL FLAX & ASSOCIATES, INC., Denver

FOOTHILL ENGINEERING CONSULTANTS, Golden

FOX & ASSOCIATES, INC., Wheat Ridge

R.D. FRITZ CONSULTING ENGINEERS, Englewood

JEROME F. GAMBA, Glenwood Springs

GAMBRELL ENGINEERING, INC., Denver

GEBAU CONSULTING STRUCTURAL ENGINEERS, Boulder

GEOTECHNICAL CONSULTANTS, INC., Colorado Springs

GEI CONSULTANTS, INC., Englewood

GEOWEST COLORADO SPRINGS, INC., Colorado Springs

GH ENGINEERING, INC., Denver

GILBERT, MEYER & SAMS, INC., Colorado Springs

GINGERY ENGINEERING CO., Denver

GOODSON & ASSOCIATES, INC., Lakewood

GORDON, GUMESON AND ASSOCIATES, INC., Englewood

GRASMICK & ASSOCIATES, Littleton

GRAY LAND CONSULTANTS, Englewood

JERRY GREENE, P.E., Gunnison

GREENHORNE & O'MARA, INC., Aurora

GREINER INC, Denver

GRONNING ENGINEERING COMPANY, Denver

GROUND ENGINEERING CONSULTANTS, Denver

HADJI & ASSOCIATES, Denver

J.R. HANEY & ASSOCIATES, Englewood

HANSON ENGINEERING, Boulder

J.R. HARRIS & COMPANY, Denver

HARZA ENGINEERING CO., Denver

HDR ENGINEERING, INC., Denver

HOLLOWAY, STAMM AND ASSOCIATES, Denver

HURST & ASSOCIATES, Boulder

HUWALDT ENGINEERS, Denver

HYDRO-TRIAD, LTD., Lakewood

IHLENFELDT-PETERSON & ASSOCIATES, Denver

IN-SITU, INC., Lakewood

INTEGRATED ENGINEERING CONSULTANTS, LTD., Aspen

INTERNATIONAL TECHNOLOGY CORP., Englewood

ISBILL ASSOCIATES, INC., Aurora

JANNEY ASSOCIATES, INC., Englewood

JEHN AND WOOD, INC., Denver

JOHNSON-VOILAND-ARCHULETA, INC, Boulder

JORGENSEN, HENDRICKSON & CLOSE, INC., Denver

KELLOGG CORPORATION, Littleton

KELLOGG ENGINEERING, INC., Golden

KILPATRICK ASSOCIATES, INC., Littleton

KIOWA ENGINEERING CORPORATION, Denver

KLP CONSULTING ENGINEERS, INC., Englewood

KNIGHT PIESOLD AND CO., Denver

KNOTT LABORATORY, Denver

E.J. KOCMAN STRUCTURAL ENGINEERING, Denver

FREDERICK H. KOHLOSS & ASSOCIATES, Wheat Ridge

KOHNERT ENGINEERING, INC., Colorado Springs

CATHERINE KRAEGER-ROVEY CONSULTING ENGINEER, Denver

JAN F. KREIDER AND ASSOCIATES, INC., Boulder

LADD ENGINEERING, Lakewood

LAMUTT & ASSOCIATES, INC., Lakewood

LANDMARK ENGINEERING, LTD., Loveland

J.E. LANGFORD & ASSOCIATES, Grand Junction

LEDERLE CONSULTING ENGINEERING, Boulder

LEIGH, SCOTT & CLEARY, INC., Denver

LENZOTTI & FULLERTON, INC., Breckenridge

LONCO, INCORPORATED, Denver

HENRY R. LOPEZ ASSOCIATES, Boulder

LOVE & ASSOCIATES, INC., Boulder

M-E ENGINEERS, INC., Lakewood

M.V.E., INC., Colorado Springs

J.E. MANZI & ASSOCIATES, INC., Littleton

MARSTON/KINNEY & ASSOCIATES, Grand Junction

MARTIN/MARTIN, Wheat Ridge

MATRIX ENGINEERS, INC., Northglenn

MCCALL-ELLINGSON & MORRILL, INC., Denver

MCCARTY ENGINEERING CONSULTANTS, Longmont

MCFALL-KONKEL & KIMBALL, INC., Denver

MCLAUGHLIN WATER ENGINEERS, LTD., Denver

R.J. MCNUTT AND ASSOCIATES, INC., Greeley

MCRAE & SHORT, INC., Greeley

MEHEEN ENGINEERING CORP., Denver

MERRICK & COMPANY, Denver

METROPOLITAN ENGINEERING, Englewood

MEURER & ASSOCIATES, INC., Lakewood

RUSSELL M. MILLER, CONSULTING ENGINEER, Timnath

MINER AND MINER, CONSULTING ENGINEERS, Greeley

S.A. MIRO, INC., Denver

JAMES M. MONTGOMERY, INC., Aurora

MORGAN & ASSOCIATES, INC., Colorado Springs

MOY-NASSAR & ASSOCIATES, Denver

MULLER ENGINEERING COMPANY, INC., Lakewood

MULLER, SIRHALL AND ASSOCIATES, Aurora

NLS ENGINEERS, Denver

NELSON ENGINEERS, Greeley

NEUJAHR & GORMAN CONSULTING ENGINEERS, Denver

NICOL HUTCHINSON, INC., Boulder

MALCOLM L. NIETZ, Colorado Springs

DON NISBET ASSOCIATES, INC., Englewood

NOLTE AND ASSOCIATES, Highlands Ranch

NORTON, UNDERWOOD AND LAMB, Greeley

OBERING, WURTH AND ASSOCIATES, Colorado Springs

OWEN ENGINEERING & MANAGEMENT CONSULTANTS, Denver

PARKING DYNAMICS, Denver

PARSONS BRINCKERHOFF QUADE & DOUGLAS, Denver

J.W. PATTERSON & ASSOCIATES, INC., Denver

PAVEMENT MANAGEMENT SYSTEMS, INC., Denver

PEAK ENGINEERING, INC., Littleton

DONALD C. PEDREYRA, Aurora

PETERSEN & ASSOCIATES, INC., Denver

A.R. PFEIFFENBERGER AND ASSOCIATES, Denver

TOM PITTS AND ASSOCIATES, Loveland

PLAINS ENGINEERING, Denver

DONALD A. PORTER, JR., Boulder

PROCESS APPLICATIONS, INC., Fort Collins

QUIMBY ENGINEERING, INC., Englewood

R & R ENGINEERS-SURVEYORS, INC., Englewood

RBD, INC., Fort Collins

R. G. CONSULTING ENGINEERS, INC., Denver

REDWINE & ASSOCIATES, INC., Aurora

REITER ENGINEERING, INC., Lakewood

LEONARD RICE CONSULTING WATER ENGINEERS, Denver

RIEGEL ASSOCIATES, INC., Denver

THE RMH GROUP, INC., Lakewood

RMS CONSULTING, INC., Aurora

ROBILLARD & ASSOCIATES, INC., Silverthorne

ROCKY MOUNTAIN CONSULTANTS, INC., Longmont

ROLLAND ENGINEERING, Grand Junction

ROTHBERG, TAMBURINI & WINSOR, INC., Denver

GEORGE V. SABOL, Brighton

J.F. SATO AND ASSOCIATES, INC., Littleton

DENNIS F. SCHICK & ASSOCIATES, Littleton

SCHNABLE ENGINEERING ASSOCIATES, Lakewood

SCOTT, COX AND ASSOCIATES, INC., Boulder

SDG INCORPORATED, Lakewood

SELLARDS & GRIGG, INC., Lakewood

SHEAR ENGINEERING CORPORATION, Fort Collins

SHIPMAN STRUCTURAL ENGINEERING, Grand Junction

SIMONS, LI & ASSOCIATES, Fort Collins

DICK SITTNER ENGINEERING, INC., Kittredge

SMH ENGINEERING, INC., Lakewood

S.H. SMITH CONSULTING ENGINEERING, Lakewood

SOIL TESTING AND ENGINEERING, Colorado Springs

STEFFEN ROBERTSON AND KIRSTEN, Lakewood

STEWART ENVIRONMENTAL CONSULTANTS, Fort Collins

JAMES H. STEWART & ASSOCIATES, Fort Collins

STRATIGEMS, Denver

STRUCTURES PLUS P.C., Lakewood

STUART ENGINEERING, Boulder

STURM + BALLARD, INC., Lakewood

THOMAS E. SUMMERLEE & ASSOCIATES, Colorado Springs

SWAISGOOD & ASSOCIATES, Evergreen

SWANSON-RINK, INC., Denver

SYSTEMS ENGINEERING CORPORATION, Colorado Springs

TARANTO, STANTON & TAGGE ENGINEERS, Fort Collins

TEC, THE ENGINEERING CO., Fort Collins

TERRACON CONSULTANTS SE, INC., Fort Collins

E.H. TIPPETS COMPANY, Wheat Ridge

TIPTON AND KALMBACH, INC., Denver

TORGERSON-YINGLING ASSOCIATES, INC., Denver

ROBERT M. TOWNER ASSOCIATES, Boulder

TRANSPLAN ASSOCIATES, INC., Boulder

TRI-CONSULTANTS, INC., Lakewood

N.V. TSIOUVARAS & ASSOCIATES, Denver

TUDOR ENGINEERING, Denver

TURNER COLLIE & BRADEN, INC., Englewood

TUTTLE APPLEGATE RINDAHL, INC., Denver

TYREE ASSOCIATES, INC., Colorado Springs

UEBLACKER ASSOCIATES, INC., Lakewood

URS CONSULTANTS, Englewood

VANEK CONSULTANTS, INC., Conifer

WALKER ENGINEERING, INC., Wheat Ridge

WALKER PARKING CONSULTANTS, Aurora

A.G. WASSENAAR, INC., Denver

WATER, WASTE AND LAND, INC., Fort Collins

OLIVER E. WATTS, Colorado Springs

WEIDMANN ENGINEERING, Longmont

RICHARD WEINGARDT CONSULTANTS, INC., Denver

WEISS CONSULTING ENGINEERS, INC., Colorado Springs

WERNSMAN ENGINEERING, Greeley

MICHAEL W. WEST & ASSOCIATES, INC., Littleton

W.W. WHEELER AND ASSOCIATES, INC., Englewood

LEIGH WHITEHEAD & ASSOCIATES, INC., Colorado Springs

JOHN W. WILLIAMS, Boulder

WILSON & COMPANY, Colorado Springs

WINDERS BARLOW & MORRISON, Denver

WOODWARD-CLYDE CONSULTANTS, Denver

WRC ENGINEERING, INC., Denver

WRIGHT WATER ENGINEERS, INC., Denver

YODER ENGINEERING CONSULTANTS, INC., Avon

YORK & ASSOCIATES, INC., Denver

CHARLES L. ZECHER, Denver

ZEILER-PENNOCK, INC., Denver

TED ZORICH AND ASSOCIATES, INC., Englewood

## INDIVIDUAL MEMBERS

Aasheim, Stephen E.

Abbott, James E.

Adams, David L.

Adams, Robert J.

Aguirre, Vukoslav E.

Ahlstrom, Scott B.

Almquist, Norman G.

Anderson, Albert E.

Andrews, A.S. "Andy"

Applegate, Charles "Mike"

Appelt, Gernot D.

Arber, Richard P.

Armstrong, Edward A.

Arndt, John D.

Arnensen, Tore O.

Attwooll, William J.

Ault Daniel V.

Austin, David E.

Baker, N. Kent

Balloffet, Armando F.

Barrett, Michael H.

Bates, Robert T.

Baur, John C.

Beardmore, Richard S.

Behrent, Robert V.

Bell, Charles E.

Bennett, Keith E.

Berge, Roger G.

Berry, Donald L.

Bishop, Harold F.

Blake, Franklin D.

Blatchley, Ronald K.

Blue, Stephen M.

Borman, James R.

Botham, Leslie H.

Boval, Lawrence S.

Bowman, Charles C.

Boyd, Daniel W.

Bredar, William L.

Brotsky, Kenneth J.

Brown, William Ken

Bruchner, Gordon W.

Brust, J.L.

Burgeson, John R.

Bushman, John K.

Butler, Gilbert L.

Carley, William Jr.

Carroll, John E.

Caspe, Gerald J.

Cator, Clinton C.

Chavez, Sol P.

Chen, Fu Hua

Cheung, Victor Y.

Church, Gary D.

Clark, Donald R.

Clay, Robert E.

Clayden, Glenn A.

Close, Steven R.

Colaizzi, Gary J.

Collins, Jack T.

Coury, Glenn E.

Cox, Garland D.

Cox, Howard S.

Croy, Don E.

Crum, Charles C.

Cunningham, John F.

Curt, Charles W.

Danzberger, Alexander H.

Davis, John F.

de Moraes, Carlos A.

Dickinson, George W., Jr.

Dickinson, George W.

Doble, Lawrence A.

Dutzi, Howard C.

East, Donald R.

Earsman, Raymond J.

Eberly, Mark J.

Eisel, Leo M.

Emmons, Robert E.

Epps, Clift M.

Erickson, Halford E.

Essigmann, Martin F.

Fairley, Thomas C.

Felsburg, Robert W.

Finch, Robert F.

Flax, Sol

Fleming, David E.

Flook, Lyman R., Jr.

Fosha, George M.

France, John W.

Frickel, Ronald N.

Fritz, Robert D.

Fujimoto, Robert I.

Fuqua, Henry H., Jr.

Gamba, Jerome F.

Gambrell, Larry J.

Gerhart, Phil C.

Greensleeve, Richard P.

Gingery, Deryl W.

Gipson, Allen H.

Goehring, Craig

Goncalves, Ricardo

Grasmick, Allen D.

Gray, Helen

Gray, T.D.

Greene, Jerry E.

Greengard, Arthur J., Jr.

Greenwald, Herbert W.

Griepentrog, Thomas E.

Gronning, Lloyd J.

Gumeson, Charles E.

Hadji, George E.

Haley, John L.

Halley, Ronald L., Jr.

Hamilton, John M.

Hamouz, William E.

Haney, Jack R.

Hanson, Roger K.

Harmon, William R.

Harral, Richard D.

Harris, James R.

Hastings, James M.

Hegg, Bob A.

Helton, Duane D.

Hendrick, John D.

Henrichsen, Karl D.

Hensen, Ronald J.

Hepworth, Richard C.

Herman, David A.

Holliday, Frank J.

Hollingsworth, Harold

Holmquist, Darrel V.

Holloway, Kirt E.

Holt, Stephen A.

Horner, Diana G.

Houdeshell, David M.

Hurst, O. Carl

Hutchinson, Ian P.G.

Huwa, Richard L.

Huwaldt, Michael H.

Isbill, H. Gregory

Janney, Jack R.

Jehn, James L.

Johnson, Theodore D.

Jones, E. Bruce

Jorgensen, IB Falk

Jubenville, David M.

Kearney, Charles T.

Kellogg, Joseph C.

Kellogg, Richard M., Jr.

Kendall, William R.

Kilpatrick, William H., III

Kimball, William R.

Knott, Albert W.

Kocman, Edward J., Jr.

Kohnert, Gerald V.

Kraeger-Rovey, Catherine

Kreider, Jan F.

Kropewnicki, Stanley J.

Ladd, Fred C.

Lamutt, Frederick R., Jr.

Langford, James E.

Laraby, James V., Jr.

Larsh, James L.

Lederle, Donald F.

Leigh, Robert E.

Lenzotti, James A.

Litton, Lester L.

Locke, Allan J.

Lopez, Henry R.

Lorah, William L.

Love, David J.

Lovejoy, Norman B.

Magnussen, Francis

Manzi, Joseph E.

Marek, Don E.

Marston, William L.

Martino, Gregory R.

Matheson, Gordon M.

Matthews, Gary D.

McCarty, John A.

McFall, James D.

McKee, Jack C.

McLaughlin, Ronald C.

McMillin, Curtis C.

McNutt, Ronald J.

McRae, Gerald B.

Meheen, H. Joe

Mehring, Clinton W.

Menhennett, Alan E.

Meurer, Douglas L.

Meurer, Gordon C.

Meyer, Edward D.

Meyers, Charles W.

Meyers, Jeffrey R.

Mighell, Edwin R.

Miller, Russell M.

Miro, Sami A.

Monahan, Donald R.

Monical, Stuart D.

Moore, William W.

Moothart, David E.

Morgan, John B.

Morrill, Gilbert B.

Muller, Larry A.

Muller, Peter J.

Nassar, Niam G.

Nelson, LaVern C.

Neujahr, Stanley G.

Newell, William H.

Nicol, Joseph C.

Nietz, Malcolm L.

Nisbet, W. Don

Norton, Thomas E.

Obering, Roland G.

Olhausen, Dale D.

Owen, William F.

Park, John D.

Patterson, John W.

Pawlak, Steven L.

Pedreyra, Donald C.

Pennock, Paul R.

Perez, Jean-Yves

Pernichele, Albert D.

Petersen, Kent L.

Petersen, Steven D.

Peterson, L. Dean

Pfeiffenberger, Andrew R.

Pitts, W. Tom

Poirot, James W.

Polhill, Dennis

Poppenga, Bernard E.

Porter, Donald A., Jr.

Priest, John E.

Proud, David J.

Quimby, H. Lavern

Rakness, Kerwin L.

Redwine, Robert B.

Rehn, Charles C.

Reich, Charles J.

Reiter, Robert C.

Renner, Robert C.

Rettberg, William A.

Rice, Leonard

Riegel, Donald L.

Rindahl, Gilbert F.

Rink, Frederick J.

Robillard, Craig F.

Robinson, Phillip I.

Rolland, Thomas D.

Roper, Gilbert L.

Rosier, Donald P.

Rothberg, Michael R.

Ruma, Sebastian A.

Rutherford, Richard A.

Ryan, James J.

Sabol, George V.

Safarian, Sargis Sako

Satchell, Thomas T.

Sato, James F.

Schaevitz, Robert C.

Schermerhorn, R. Stephen

Schick, Dennis F.

Schmeling, Karl H.

Schmid, Michael N.

Scott, William G.

Sellards, George D.

Serencko, George J.

Shear, Brian W.

Shepherd, Thomas A.

Shipman, Joseph M.

Siano, James A.

Signs, Cheryl L.

Simons, Kenneth B.

Sittner, Richard W.

Smith, Chester C.

Smith, Merritt E.

Smith, Stanley H.

Smith, Zachary A.

Sovern, Douglas T.

Sparlin, Richard F.

Spicer, Jerry L.

Steele, Richard E.

Stevens, Ronald L.

Stewart, David R.

Stewart, James H.

Storms, Lawrence D.

Stuart, E. John, Jr.

Sturdevant, Bruce L.

Sturm, Richard D.

Suedkamp, Richard J.

Summerlee, Thomas E.

Swaisgood, James R.

Swanson, James R.

Taggart, William C.

Tallarico, Robert A.

Taranto, Donald N.

Tebbens, Carl A.

Ten Eyck, Gregg S.

Thompson, Robert W.

Tippets, Elvis H.

Tobler, Leroy E.

Tolle, William A.

Toren, Ralph L.

Torgenson, Verlin G.

Towner, Robert M.

Tsiouvaras, Nicholas V.

Tyree, Perry C.

Ueblacker, Horst

Ullman, Thomas F.

Underwood, George M.

Vanek, Michael D.

Walker, Bruce A.

Walker, Robert J.

Wassenaar, Allen G.

Watts, Oliver E.

Weidmann, Victor H.

Weingardt, Richard G.

Weiss, Gerald J.

Wernsman, Stephen C.

West, Michael W.

Whitehead, Leigh

Wiechman, Merton L.

Williams, John

Williams, John W.

Wilson, William W.

Windolph, Gary R.

Worrel, Richard S.

Wray, H. Lynn

Wright, Kenneth R.

Yester, Robert J.

Yoder, David L.

York, Jerry J.

Zecher, Charles L.

Zeiler, Jack C.

Zimmer, Eugene G.

Zorich, Theodore M.

## CECC LIFE MEMBERS

Becker, Ralph W.

Johnson, Clifford

Ketchum, Milo S.

Koch, Paul H.

Kolstad, C. Kenneth

Konkel, James

Kraettli, H. Jack

Langebartel, Walter S.

McDonald, Gordon A.

Meurer, Malcolm R.

Nedell, Robert S.

Phillips, Orley O.

Preszler, Donald L.

Wilson, Morey S.

*George and Virginia Koonsman with guest at CECC Awards Banquet (late 1970's).*

**THE 10 DISCIPLINES MOST OFTEN PROVIDED BY CECC MEMBER FIRMS**

The chart shows percentage of firms offering the service noted.  Many firms provide more than one engineering discipline.

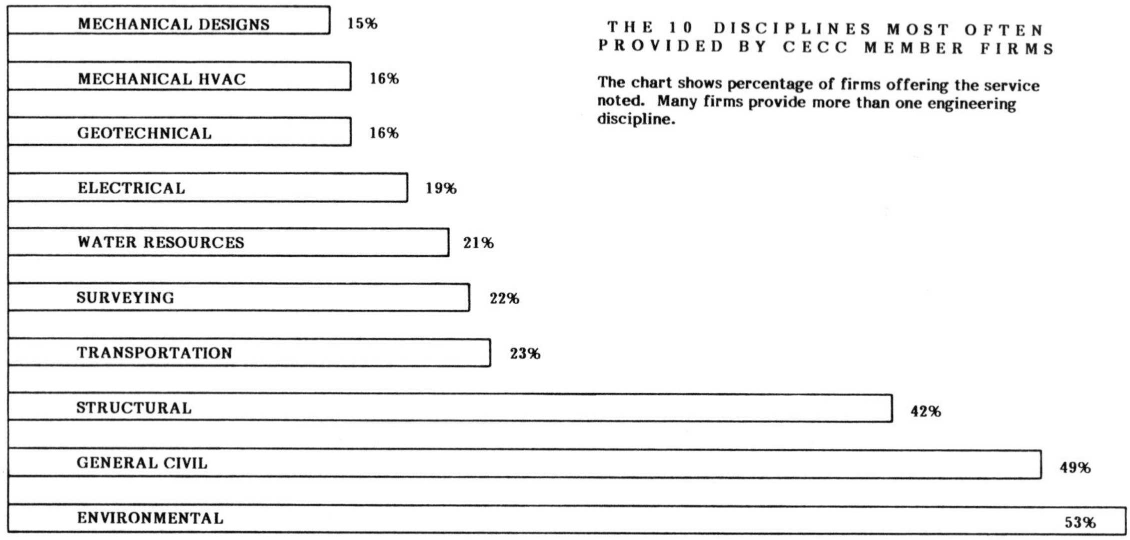

# Statistics from 1988 Member Survey

*From a November 10, 1988 survey of CECC member firms, seventy-three percent (73%) of the firms responding predict 1989 will be better for them. They anticipate that their gross billings will be nearly eighteen percent (18%) higher than in 1988.*

*The reason most often given for their optimism is that they look for an increase in the amount of design work for out-of-state projects.*

*Because recent years of a sluggish Colorado economy have had a harsh negative impact on the state's design and construction industries, it is not surprising the survey revealed that seventy-nine percent (79%) of all Colorado engineering firms do design work on out-of-state projects. In fact, thirty percent (30%) receive over fifty percent (50%) of their fees from out of state. Seventeen percent (17%) of all Colorado firms do most of their engineering work nationwide and not in Colorado.*

*If predictions are right, consulting engineers in the state will have gross billings in 1989 of $267 million, $70 million of these billings will be brought into Colorado*

*from work performed for out-of-state clients on out-of-state projects. CEC member firms currently employ 4100 persons and projections are that the number will increase to 4600 by the end of next year. Forty-five percent (45%) of all firms responding say they have CADD equipment in-house.*

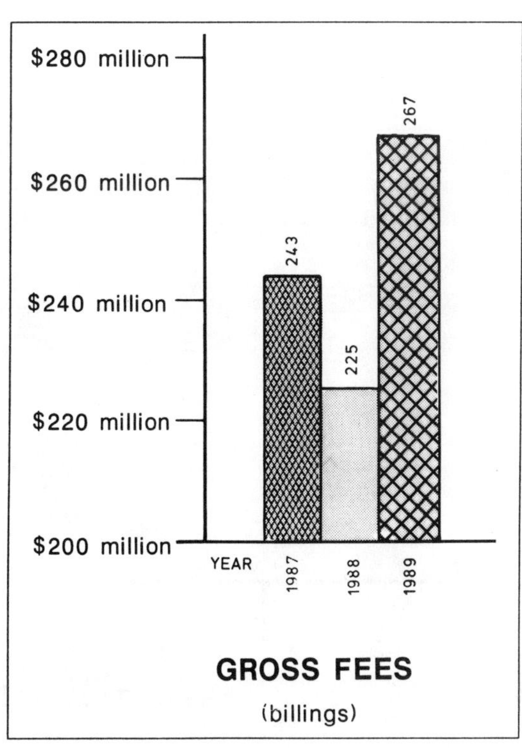

NUMBER OF EMPLOYEES
60
50
40
30
20
10
0
YEAR
1987 1988 1989
ALL DISCIPLINES
20.2 21.2 24.2
1987 1988 1989
WATER RESOURCES
10.2 9.4 10.4
1987 1988 1989
CIVIL
28.4 31.3 36.1
1987 1988 1989
MECH AND ELEC
14.2 15.7 17.0
1987 1988 1989
GEOTECHNICAL
51.0 54.0 57.0
1987 1988 1989
STRUCTURAL
4.6 4.4 5.1
AVERAGE NUMBER OF EMPLOYEES PER FIRM
(1989 FIGURES ARE PROJECTIONS)

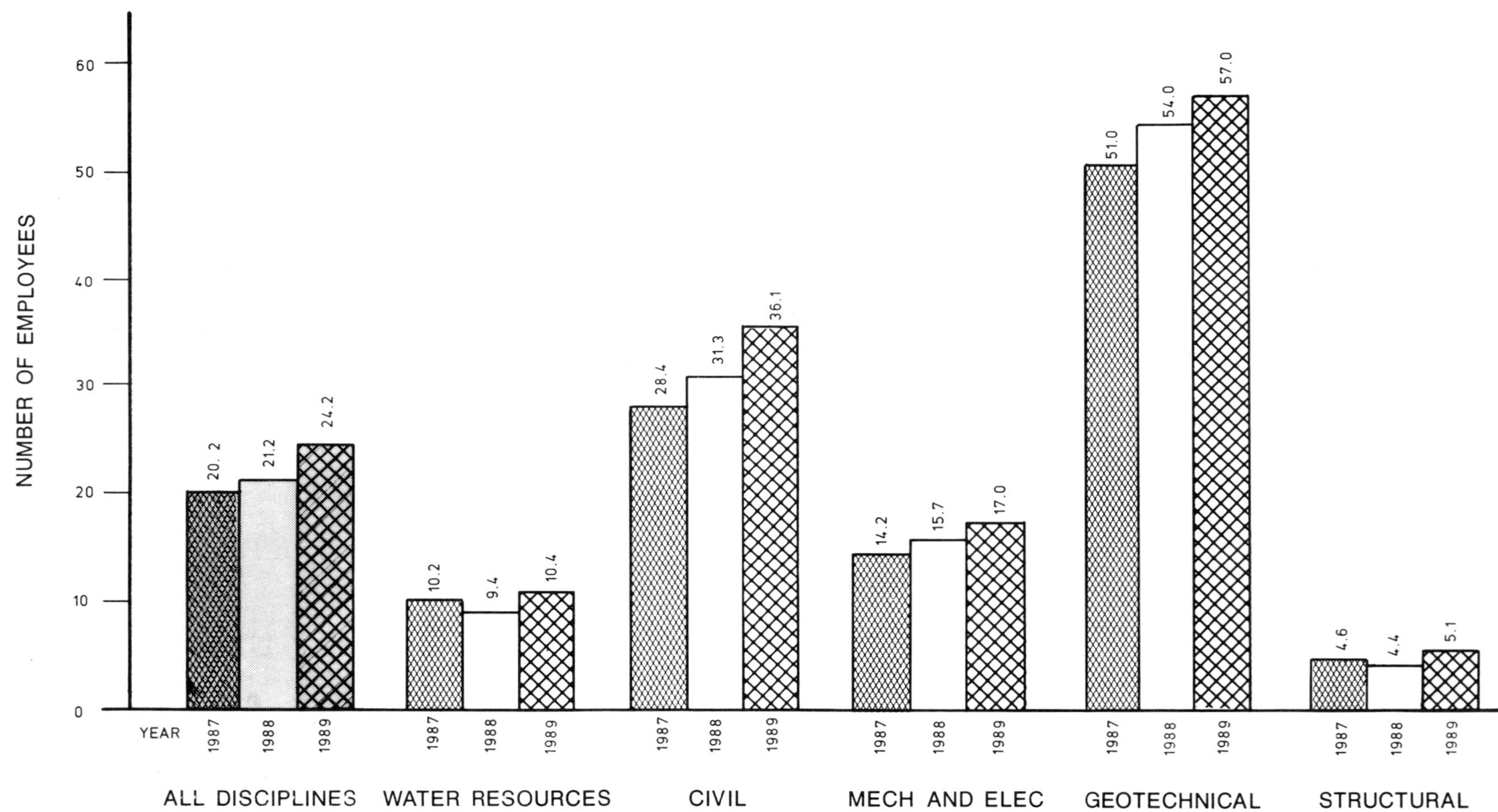

158

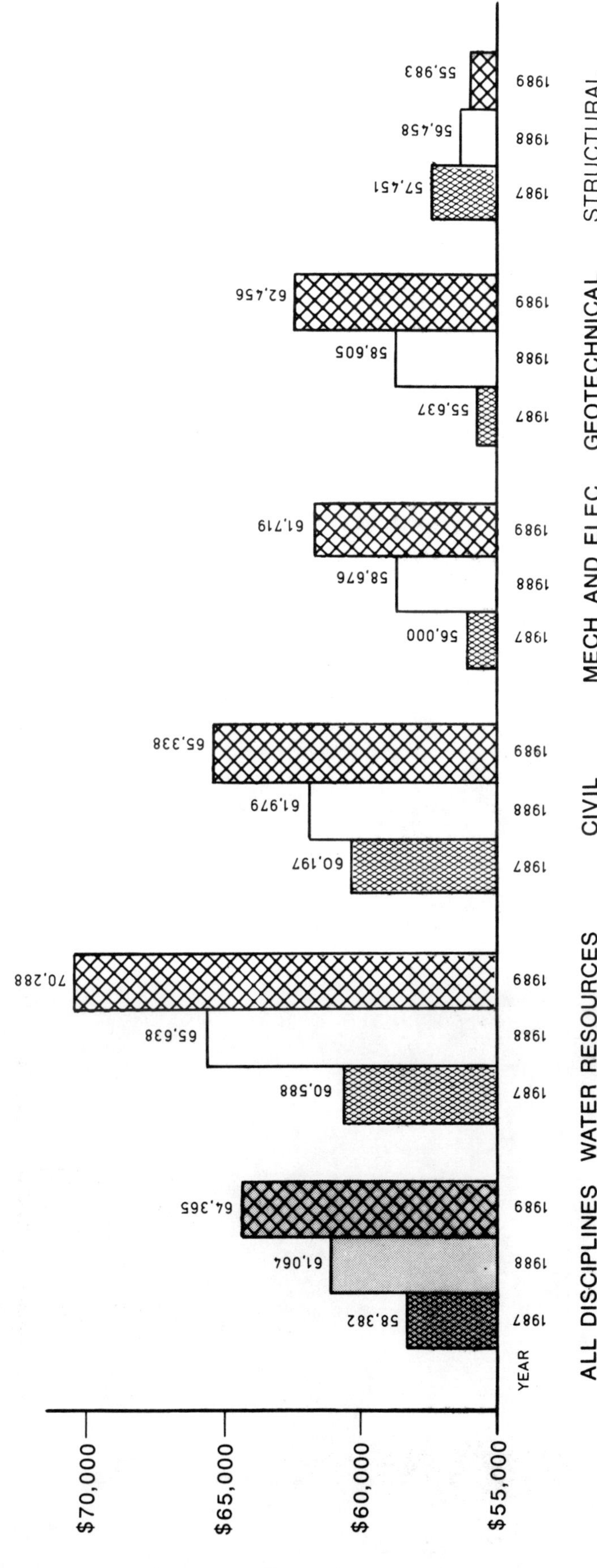


## AVERAGE GROSS BILLINGS PER EMPLOYEE

(1989 FIGURES ARE PROJECTIONS)

# Code of Ethics

With high regard for the engineering profession and recognizing in the Code of Ethics a set of dynamic principles to guide his services to his fellowmen, and with full knowledge of the responsibility of consulting engineers to safeguard health, safety, and public welfare, a member of Consulting Engineers Council of Colorado:

A.  Brings credit, honor and dignity of the engineering profession to his dealing with clients, other engineers, and the public.

B.  Acts for his clients as a faithful agent or trustee and accepts remuneration only in accordance with his stated charges for services rendered.

C.  Exchanges non-confidential engineering information with other engineers, students, and engineering press, encourages the public knowledge of engineering, and issues proper credit for engineering work.

D.  Solicits engineering work assignments according to high professional standards, without offering commissions or using undue influence.

E.  Upholds and promotes the principle of selection of consulting engineers for assignments on the basis of qualifications including training, skill, experience, personnel, work loads, and availability.

F.  Associates as a consulting engineer only with engineers and other professionals who conform to ethical practices.

G.  Shall be familiar with, and abide by, the ACEC Professional and Ethical Conduct Guidelines.

H.  Continue their professional development throughout their career, and shall provide opportunities for the development of those under their supervision.

# Index

# T

Tacoma Hydroelectric Project - *77*
Tallarico, Robert - *55, 56*
Ten Eyck, Gregg - *54, 121*
Thompson, Bob - *108*
Thorson, Edward - *13, 33, 38, 40, 41, 108*
Todd, Norman - *14*
Tracy, John - *14, 21, 86*
Trego, William - *14*

# U

Union Colony Civic Center - *77*
University of Colorado
    Engineering Science Center - *77*

# V

Vail Pedestrian Bridge - *77*
Vail Public Library - *78*
Valley Highway - *22*

# W

Waggoner, Eugene - *20, 21, 27, 86, 135*
Wallace, George - *53, 121*
Wallace, Jeremy - *53*
Wattenberg, David - *121*
Webb, Peter - *57*
Weers, Arthur - *14*
Weidmann, Victor - *137*
Weingardt, Evelyn - *42, 43*
Weingardt, Richard - *8, 12, 47, 51-54, 56
    57, 101*
Well Pump Station A-5A
    (Arapahoe County) - *78*
White, Kenneth - *14, 27*
Whittlesey, Robert - *14*
Williams Fork Dam - *60*
Williams, George - *30*
Wilson, Maurice - *14*
Windman, Arnie - *45*
Wirth, Tim - *41*
World War II - *13*
Wright, Kenneth - *25, 27-30, 34, 43, 90, 91*

# Z

Zook, Fred - 46
Zuni Facility - 78

# CONSULTING ENGINEERS ASSOCIATION of COLORADO
# —BULLETIN—

| EDITOR | APRIL | ASSISTANT |
| --- | --- | --- |
| Malcolm R. Meurer | 1965 | Harvey A. Kadish |


## Gene Takes Office Next Month As CEC President Elect

EUGENE WAGGONER

Eugene Waggoner will be installed next month as President-Elect of the Consulting Engineers Association, USA, at the Second National Convention in St. Louis, May 19-21.

Waggoner, 52, widely respected throughout the Rocky Mountain region for his engineering ability, integrity and experience, is well-known throughout the nation for his continuous efforts to improve the engineering profession.

In his contributions to engineering as a practitioner, as a thought leader and opinion shaper, and in his efforts in behalf of his community and nation, Gene has reflected great credit upon himself and his profession.

Following his studies at the University of Southern California, where he earned his bachelors of Science degree in 1937 and his masters of Science in 1939; Gene has devoted more than 27 years to his profession.

E. VERNON KONKEL, our other man in the national CEC ADMINISTRATION. Colorado was represented at the top level last year by Vern. He continues this year as a national vice president.

### A CEAC PERSONALITY

KENNETH R. WRIGHT, president of Wright Water Engineers, Denver, is serving his second term as chairman of the CEAC Public Relations committee. Ken, a holder of a bachelors and masters degree from the University of Wisconsin, has headed his firm for six years.

### NAMES IN THE NEWS

James H. Konkel Consulting Engineers, now to be known as McFall and Konkel Consulting Engineers. James D. McFall becomes a partner.

Wright-McLaughlin Engineers, formed by Kenneth R. Wright and Ronald C. McLaughlin. Wright continues as principal of Wright Water Engineers.

### A CEAC PERSONALITY

JOHN E. BUNTS, structural engineer practicing in Colorado Springs. Currently president of CEAC, Jack is a graduate of the University of Colorado. His CEAC duties involve numerous trips to Denver and other parts of the state and nation.

### OVERSEAS WORK?

A weekly magazine, INTERNATIONAL COMMERCE, published by the U.S. Department of Commerce, lists planned construction projects overseas. The U.S.D.C. office in Denver has a complete library of back issues, or subscriptions can be obtained for $16 a year from the Denver office.

Attention is called to 50 projects approved by the Governing Council of the U.N., listed in the August 3 edition.